First Law
Of Physics,
Let There Be
Light

First Law Of Physics, Let There Be Light

Miles Pelton

To order additional copies of this book, contact:
Xlibris LLC
1-888-795-4274
www.Xlibris.com
Orders@Xlibris.com
142381

CONTENTS

PART I

Fundamental Physics Principles

PART II

Examples of Creation Physics Power

PART I

Fundamental Physics Principles

Introduction I

Merriam-Webster defines ***physics*** as "a science that deals with matter and energy and their interactions." It defines *religion* as "(1) the service and worship of God or the supernatural (2) commitment or devotion to religious faith or observance." It defines *creation* as "(1): the act of creating especially: the act of bringing the world into ordered existence. (2): the act of making or producing. (3): something that is created as: (a) the world. (b): creatures. (c): a work of art. (d): a new or striking article of clothing."

Going with the generally accepted conclusion that matter and energy are the ingredients of creation, it follows that physics involves the study of the fundamentals of creation, which mandates that the processes fundamental to creation be understood and accepted to form the foundation for all other physics matters. Therein lies a problem. The researchers and academia that have made physics their life's mission follow laws and principles that establish an anticreation mind-set.

After decades of thought and debate trying to come to agreement on the methods and processes involved in powering creation, there evolved a theory that all behavior in the universe was powered by the perpetual motion of the particles of matter of which all things were made. That kinetic theory of matter was adopted by physicists without proof as was the conservation of energy law. The anticreation or anticreator mind-set mirrors the Satan-driven behavior documented by the scriptures wherein anticreator beliefs were attributed to Satan, who was apparently accepted as normal.

The purpose of this writing is to try to bring awareness of that situation so physicists can concentrate on letting the facts in evidence lead where they will and religions can concentrate on worshipping God. The first order is to understand that a biased mind-set exists and interferes with correct understanding.

The prevailing science-religion feud began when the Hebrew religion recognized creation as the foundation of their belief system that is defined and set as the opening of the written word adopted as biblical scripture. The feud was set when the community of scholars studying physics found the expressed *religious beliefs* relative to creation were deemed not technically correct and therefore not acceptable as physics principles. There can be disagreement over beliefs, but that doesn't rule out the fact there was a creation event. The problem arose because the contested argument is framed on *beliefs formed* relative to the creation event and not on the facts that define the event.

The beliefs that formed the Hebrew religion were and are based upon one God, creator of all things visible and invisible, which puts belief in creation at the forefront of the Hebrew and consequently Christian and Muslim beliefs. The Bible written to document Judeo-Christian beliefs opens with an outline of beliefs relative to creation processes. Those are an outline of reasonable beliefs that do no harm but provide an understandable nontechnical summation of the matter that instills hope, faith, and optimism in believers. They were presented as an argument to be believed.

There are many that do not subscribe to those religious beliefs especially in regard to creation being engineered and powered by a divine being, among whom are the community of science scholars and the academia that study creation to extract and teach physics principles inherent in the processes of creation. As a consequence, whether willfully or unwittingly, the kinetic theory of matter and the law of conservation of energy offered an alternative theory. The conspiracy that proceeded from those beliefs is profound and extends to hold dominion over most of secular academia, textbooks, and the civil government worldwide. It is a conspiracy that demands a biased mind-set that prevents an understanding of fundamental physics and has produced a growing collection of distorted beliefs regarding physics principles.

Revelation of the flawed physics fundamentals as explained herein developed over a period of near twenty years and began with a challenge of a younger brother, a former Southern Baptist minister, to explain gravity. A stock textbook, professional-engineer answer did not satisfy his inquisitive mind. He sent a sketch to illustrate why the textbook explanation was defective and asked for reconciliation of the difference. He died unexpectedly before a response could be made, but the illustration he had provided begged for an answer.

Engaged in mowing the grass of the ten-acre soybean field on which my parish decided to build its church afforded considerable thinking time. Over time, a realization developed that the weeds and plants had a basic "if this, then do that" inherent intelligence since they knew when to grow and when to bloom. This intelligence was especially noticeable with the purple heather-like and yellow oxalis-like weeds that bloom to announce Easter. That conclusion led to the question, could the matter fundamental to creation have *intelligence* by which gravity could be explained? The challenge to explain gravity was ever present.

On an occasion while attending late-afternoon church service where the tendency for mental wandering is ever present, the sun shone near horizontally so that the Altar candles cast shadows on the wall behind. On one such occasion, it registered that while the shadow of the candle was clearly cast, the candle flame, though clearly visible, did not cast a shadow. For some reason, on this occasion, it was instantly apparent that the flame, although clearly visible, had no physical presence. That observation culminated in the experiment documented in a subsequent chapter and realization that the candle flame was a glow produced by the energy that bound the atoms of the elements involved in producing the flame.

The challenge over gravity centered on trying to explain Galileo's Leaning Tower of Pisa findings. How did gravity know to apply force on objects on the basis of density and not physical size or type of material? How did gravity, acting over distance with no intervening material medium, know how to adjust force to compensate for distance? How did gravity know the density of the material being attracted? These were questions that were a mystery even to Sir Isaac Newton, the master of research and study of gravity.

Eventually it dawned that there could be only one explanation: the energy that produces the force of gravity has to exist inherent in the fundamental particles of which all matter is made. From that conclusion grew consideration for the ancient Chinese philosophy of yin-yang forces. That led to the conclusion that the mystery of gravity could be explained by acknowledging that elementary particles with absolute yin-yang (self-affinity plus self-repelling) forces are fundamental to the creation of matter. The rest of the explanation is detailed in a subsequent chapter.

With gravity explained, only the mysteries of electromagnetic force, weak force, and strong force remained to be solved. (Einstein conceded to four unsolved mysteries in physics: gravity, electromagnetic force, weak force, and strong force). Years of experience with electric generation and the understanding revealed by the illuminated candle led to a realization that the remaining outstanding mysteries all dealt with a common phenomenon, the relationship that exists between protons and electrons wherein protons are the controlling factor. That led to the conclusions explained in subsequent chapters.

Please do not misunderstand. This presentation is not fiction or a novel. The fundamental elements of creation are not visible and incapable of being effectively measured yet an understanding of the physics principles fundamental to creation is necessary to develop relevant, realistic physics and religion beliefs and principles. The feuding, biased, and flawed beliefs of both physics and religious teachers make it difficult for maturing minds, especially, to develop sound beliefs. That is the motivation behind this effort.

CHAPTER ONE

The Doorway to Understanding

It is important to realize, in the logic inherent in the vernacular of olden times, that the horse goes before the cart; in this case since protons, neutrons, and electrons that are constituent components of atoms that are themselves elemental to the formation of all matter, protons, neutrons, and electrons had to have existed before the formation of any matter. Going one step further, the components of which protons, neutrons, and electrons are made had to have existed before those particles elemental to atoms could have been made. Since the materials elemental to the creation of protons, neutrons, and electrons are not visible or possessing discernible physical properties, identification and definition is dependent upon consideration of what properties would be required to produce the characteristics and behavior inherent in protons, neutrons, and electrons. At this point, old-time logic says binding of elementary particles is a necessary ingredient in the formation of atoms and molecules and that the energy to produce binding is a fundamental component of creation.

As documented in textbooks and research literature, it is recognized that neutrinos or neutrino-sized particles are elementary to the creation of neutrons and thereby all matter. That gives rise to the question of how one type of particle can produce matter with a physical presence and the power to function. There is a logical explanation that was postulated in ancient Chinese culture. There is a two-component charge on each elementary particle. The basic charge component imparts to each elementary particle the power to attract the like charge of another like particle. The second charge component imparts to each elementary particle the power to repel the like charge of another like particle. Then the attraction charge has the power to attract the repelling charge. The ancient Chinese recognized the attraction and repulsion forces involved in creating matter and called them yin-yang. The attraction-repulsion charge imparts the power needed for elementary particles to assemble while the repelling charge acts to create physical presence by preventing the charges from being pulled into a singularity. (A more definitive explanation of the energy charges associated with elementary particles is provided in the blog included in Part II, entitled Trinity Particles.)

An anticreation bias permeates physics in the form of theories and laws that maintain that a continuing process of perpetual motion within elementary particles provides the energy to power the behavior of matter, thereby negating the need for a source to supply empowering energy. In other words, the belief prevails that creation is self-powered, negating the requirement for a continuing supply of energy, thereby disparaging creation and the concept of creation. The process physicists believe is documented as the kinetic theory of matter that is implemented by the conservation of energy law. Those beliefs were developed and adopted without being proven, without explanation of how the alleged motion is converted to binding energy or how and what powers the perpetual motion.

It should be understood that existing test instruments are unable to sense and measure the force producing potential of the attraction charge impressed on elementary particles because the charge is not an electrical charge and has no proton reference. Without a proton reference against which to measure a difference of potential, the charge cannot be detected or measured with electric-charge-measuring equipment. (The repelling charge component is isolated to form electricity so it can be detected and measured with instruments that have been developed for the purpose.) The existence of a charge that powers attraction is obvious by the fact that the fundamental particles produce attraction force, otherwise known as gravity, but gravitons, the quanta energy particle of the attraction energy component, remain undetected for the reason explained. It is worthy of note that for other processes such as those researched by Coulomb, the observation of a force was accepted as proof that a force-powering charge existed, yet for some reason, the existence of gravitational force is not accepted as proof that an attraction charge exists. Perhaps because to acknowledge that would render the law of conservation of energy invalid.

The repelling and the attraction component of the energy that conveys power to elementary particles interacts with the attraction energy component on other like-charged particles only. Each charge unit can interact with any number of other units; interaction is not one-on-one. Repelling charges repel the repelling charge on other like elementary particles. Attracting charges attract the attracting charge on other like elementary particles. The attracting charge and the repelling charge attract each other with the repelling charge encapsulating the attracting charge to form the elementary particle. It is the attraction component charge that powers the force called gravity or gravitational attraction. Each elementary particle carries one unit of attracting and one unit of repelling charge.

Protons transform the attraction and repulsion charges carried by the elementary particles of which the proton is made into lines of force with one line of force emitted by each proton. A more detailed explanation of the proton behavior is provided in a subsequent chapter. The lines of force emitted by protons consist of a line of streaming protons that are packets of 1,840 attraction energy units that are encased by a stream of electrons (repelling charges) that are made to spiral in the process of being emitted. After emission by a proton the repelling component charges have a reference to, they are attracted by the proton charge center and are then known as electrons. The lines of force emitted by protons establish the bonds that bind atoms and assemblies of atoms. When the repelling charges are stripped from the lines of force emitted by protons and impressed upon an insulated isolated system of conductors, as is the process of an electric generator, they make electricity. (That process is explained in greater detail in a subsequent chapter.)

One thousand eight hundred forty elemental attraction charges (there is some disagreement on the exact number) are assembled into packets called photons that are encased into the lines of force emitted by protons. In the course of being emitted by protons, the photons are encased within the repelling component charges (i.e., electrons). The self-attracting photons constrict to apply a bonding force that binds atoms and assemblies of atoms. When the encasing electron charge of a line of force is fragmented, as when exposed to excessive heat or physical trauma, the encased photons radiate from that point at the speed of light; this phenomenon is called light. The energy carried by the radiated photons produces the effects called light and heat, which are explained in greater detail in a subsequent chapter. These are energy forms and forces that are fundamental to physics, and the processes involved are absolutes throughout the universe.

It is important to recognize that the attraction and repulsion energy components on the elementary particles are sustained as constant because the presence energy charges implanted on elementary particles are sustained. Translated, that means energy expended to produce binding force is constantly replenished, and since elementary particles are components of all matter, they are the carriers of an absolutely constant and constantly maintained supply of energy. The process by which the charge maintained on elementary particles is generated is explained in the following chapter.

The conservation of energy theory, adopted as a law, is filled with unproven, erroneous anticreation claims. Essentially the conservation of energy theory claims that atoms and thereby mass are convertible to energy, and that energy is convertible to mass, a perfect machine that does not require energy for power. The principle flaw stems from the anticreation denial that energy is consumed in powering the creation and sustenance of atom-forming bonds, which in turn makes it impossible to explain mass in the terms defined by Newton.

Mass as defined by Newton and now called rest mass should reflect the number of elementary particles in an assembly. In other words, the mass of a neutron or a proton or an atom should reflect the number of elementary particles in the assembly. For example, with each elementary particle charged with one unit of attraction energy, an assembly of 1,840 elementary particles would have a center of gravity with a 1,840-unit attraction intensity. However, current physics beliefs and principles do not recognize the stated correlation between mass and gravity. Instead, driven by the conservation of energy law, the theory of relativity, and other hypothetical beliefs like the equivalency theory, mass has been defined as "the resistance of an assembly to acceleration by an outside force." In other words, the force required to produce acceleration is taken as the mass of the object because, by current reasoning, that measurement reflects the potential of that mass when in motion to produce force, thereby the energy inherent in that object. As a consequence, the term *mass* currently defines the kinetic energy of an object in motion and has no relevance to the number of elementary particles or the energy involved in creating the object or the objects power to produce gravitational force. It is a hypothetical figure with no relevance to reality. It takes physics into a world of make-believe supported by a system of mathematics developed to justify those hypothetical beliefs.

While the measurements obtained by the process outlined may have value for calculating other interactions between bodies or the kinetic energy potential of a body, those calculations are not a reflection of the mass of the object in the terms explained by Newton, and they are not compatible to understanding fundamental physics especially gravitational attraction, mass, the energy involved in mass creation (fusion) and mass destruction (fission), or the energy that produces light, heat, electricity, and magnetism. Those processes are explained more fully in subsequent chapters.

The term *mass* has also been twisted by the belief in equivalence embodied in the conservation of energy law with the fallacious assertion that mass and energy in any object are interchangeable. In fundamental physics, the term *mass* is a function of the number of elementary particles employed in making the object where the input of energy to bind the elementary particles is a requisite and the removal of elementary particles involves the release of the energy that had been applied to bond those particles. The mass of an object is increased or decreased by changing the number of elementary particles, not by converting the mass of elementary particles to energy. The addition of particles to an object requires that energy be expended to effect and sustain the force that bonds the added elementary particles, which means that the energy expended to make and sustain a bond must be replenished if the assembly is to remain viable. When the number of particles is reduced, the amount of energy required to apply bonding force is reduced. Since the bonding force is produced by the photons inherent in the line of force producing the bond, when a line of force is fragmented, the photons are released to radiate; ergo the energy carried by the radiated photons is lost from that assembly. Further, if or when another elementary particle is added to the assembly, the energy to establish and sustain the requisite bond is inherent in the line of force supplied by a nuclei proton. It is not supplied by converting mass (elementary particles) to energy. The energy lost by any previous loss of mass was radiated as photons and is no longer available to power the formation of new bonds.

It has not been established how much energy it took to create elementary particles if any or that any elementary particles are or can be destroyed. Bear in mind there is only one elementary particle, and it is made of one charge of attraction, one charge of repelling energy and one charge of attraction-repelling affinity energy. Neutrons and protons are an assembly of elementary particles bound by the self-attracting energy component, which has bonds that cannot be fragmented. (The bond between the attraction and repulsion components of presence energy can be separated/isolated as in the process of making protons or generating electricity.) All other assemblies (atoms and molecules) are made of protons and neutrons bound with the energy contained in the lines of force emitted by protons, which has lines of force that are fragile. Therefore, with conversion of elementary particles excluded as factors, any change of mass or energy of an object involves only the making and breaking of interparticle bonds. Simply stated, it is not mass (matter) that is converted to energy, and it is not energy that is converted to matter (mass).

Whether elementary particles exist as independent assemblies or are assembled as in neutrons and protons and thence atoms and assemblies of atoms (molecules), the assembly acquires a center of gravity and a center of proton power, equal in intensity to the number of elementary particles in the assembly. Those centers of gravity and proton power establish a charge field that surrounds the assembly, which has a charge that exerts a force on the charges on the elementary particles in other objects. The attraction charge (gravitational attraction) is detectable and gives indication of the number of elementary particles in the assembly, its mass. However, the force measured must be adjusted for the reduction in intensity due to the distance-squared rule, the reduction in energy intensity due to heat, and forces applied by other massive objects.

Scales used to determine weight—for example, of a person on earth—measure the vector sum of the forces applied by the earth, the moon, the sun, the Milky Way and the individual being weighed. The result is sufficiently precise for the purpose of determining weight, but weight is not a precise measurement of the mass of the object. As a consequence, the term *mass*, as currently understood, is and has led to flawed beliefs and principles especially calculations made on the basis of data provided by the periodic table of the elements.

The great minds that cooperated in developing the atomic and nuclear bombs relied on the periodic table of the elements to prove their theory on the source of the energy released by fission and fusion reactions. The periodic table of the elements was developed by a chemist; based principally on chemical reactions, it established the atomic weight for each of the known elements of matter. Working principally with a spectroscope, the apparent weight (mass) of an atom of the various elements was calculated. Technology does not exist for the direct measurement of the energy component that produces the force of gravity and thereby mass. Without instruments capable of measuring mass, the mass of elementary particles and the various element assemblies were established by comparing the reaction and extent of reaction when the object being studied were forced to traverse magnetic and electric force fields. The results were established by imprecise methods. Without elaborating, the data documented in the periodic table of elements is suspect because it was the intensity of the attraction charge of the assembly not the number of elementary particles that was determined. Furthermore, the intensity of the attraction charge diminishes as distance is squared, thereby requiring consideration especially when working with large atoms.

The process involved in developing the periodic table of the elements tried to establish the weight (mass) of nuclei and the elementary particles that make nuclei. Working principally with the hydrogen atom, which is made of one proton without its electron and one neutron, the effect of charge forces acting on the target particle forced to traverse a magnetic and electric force fields was translated into the particle's mass energy. Bear in mind that the effect established by this measuring process was the apparent kinetic energy of the target, not the rest mass or the intensity of the attraction producing energy. Even if precisely accurate, the results did not establish the energy that was expended to make the target particle. In other words, the results did not precisely establish the target's atomic weight.

When tasked to explain where the energy released by nuclear reactions comes from, physicists turned to the imprecise periodic table of the elements that shows kinetic energy potential, not the energy inherent in the target particle. The data published in the periodic table shows that atomic weight units determined for a single proton and a single neutron, when added cumulatively based upon their number in a specific atom, was greater for some elements (those less than the size of the lead atom) than the apparent weight of the assembled atom. That difference was taken to mean that the energy released by nuclear fission and fusion reactions was explained by the apparent deficit shown by the periodic table. Those conclusions showed that energy was released by fusion reactions even though fusion requires the expenditure of energy to power the establishment of the new bonds.

Faced with the need to show compliance with the law of conservation of energy, the apparent contradiction revealed by the periodic table was accepted because it was supported by the periodic table. From that, without justification or explanation, a conclusion was drawn that the difference in apparent mass confirmed that it was the conversion of mass to energy that powered nuclear reactions. Also, since the energy released by combination fission and fusion reactions was significantly greater than a fission reaction, it was concluded that fusion released vast quantities of energy. It appears no one noticed or wanted it to be noticed not only that the precision of measurements published in the periodic table was less than precise but that the formation of bonds consumes, and does not release, energy. Also, no one seems to realize that fissionable material was produced by the fusion reaction and that it succumbed to fission in the combined process.

The point is, the only things lost or gained in destroying or creating mass, even in atomic or nuclear reactions, are the bonds that bind the particles to form atoms (mass), and the energy to form and sustain bonds is supplied through the elementary particles of which the matter is made. Mass (meaning elementary particles) is not destroyed or produced by nuclear reactions, and bonds are not made and sustained with energy derived through conversion of elementary particles.

It is the attraction component of the fundamental particles that produces the fundamental force called gravity or gravitational attraction (GA). The repelling charge component of the fundamental particles is what is now called an electron. It is the combination of attracting and repelling forces, called yin-yang forces by old-time Chinese philosophy, that powers the creation of matter and mass and that supplies the power to sustain the functioning of creation, which includes solar systems, galaxies, and the universe itself. The number of elementary particles existing throughout the universe is mind-bogglingly significant and, as such, constitutes a corresponding mind-boggling energy potential. While beyond the ability to establish precisely what physics buffs call empirical evidence, it can be concluded that the energy potential inherent in that mass of elementary particles will be sustained as long as the mass of particles exists. It is suggested that it is the energy potential inherent in that mass of elementary particles that supplies the energy to power their assembly to create matter (mass). The processes involved in such an energy-generating system are explained in the next chapter.

The intensity of the force applied by one assembly on another varies for specific reasons. One is that the intensity and thereby the ability to produce force decreases inversely as the square of the distance from the center of the assembly, the center of gravity, or the center of proton power. Second is that proton power intensity is depleted by and equal to the magnitude of heat energy that is taken up by the proton center. Third, the intensity of the power of a center of gravity or a center of proton power equals the number of elementary particles contributing to the assembly in the case of gravity, or the number of protons contributing to a proton power center of an assembly.

As elementary particles assemble in response to gravitational attraction (GA), the resulting assembly acquires attracting power or intensity equal to the number of assembled particles, which acts from the vector center of the assembly, its center of gravity. While possessing attraction intensity equal to the sum of assembled particles, a center of gravity is limited to attracting individual elementary particles only. That is, the center of gravity of a body attracts the individual particles of another body. The number of elementary particles in an attracted assembly establishes the number of lines of force that can be applied by the attracting body. Since the particles of the attracted body are themselves bonded, the force applied to the body corresponds to the total of the force applied to the individual particles. That explains the findings of Galileo's Leaning Tower of Pisa research.

Attraction bonds will stretch to infinity but will never break, they just weaken with distance and fade away like old soldiers that never die. The energy that powers gravity is supplied from a singularity and is therefore instantaneous-acting presence energy. The energy that powers gravity no physical presence and is detectable only by the force of gravity or gravitational attraction. Attraction between elementary particles is directional and instantaneous but the force-producing strength of the charge is subject to reduction due to the inverse distance-squared rule even though the reaction is instantaneous. Furthermore, the power of an assembly center of gravity is applied in attracting elementary particles in order of distance from its center of gravity until the total available power intensity of that center of gravity is committed, and that order begins with the binding of the particles of which the assembly is composed.

While self-binding is the first order in every instance, the inverse squared law means the center of gravity of each assembly has more power intensity than is committed to self-binding and the remaining power is available to attract elementary particles of other assemblies. For example, because of the inverse squared distance factor, the earth has enough power after self-binding to attract and hold the moon in orbit. It does so by attracting the individual particles of which the moon is made that are, in turn, being held as an assembly by self-binding. In the same manner, after self-binding, the moon has sufficient attraction power remaining to attract elementary particles on the earth as evidenced by tidal action on the earth's oceans, and which illustrates that the moon's center of gravity attracts the elementary particles of which the earth is made. Because the elementary particles are bound to make molecules, water molecules are moved by the GA of the moon to produce tides on the earth's waterways. Recognize that GA acts line of sight to apply force to elementary particles individual as evidenced by tides produced by the moon on the far side of the earth.

Gravity (or GA) is a universe wide absolute fundamental to creation. Gravity is, without exception, the same on any assembly or body in the universe, and in that context, solar systems, galaxies, and the universe each constitute an assembly. Further, each identifiable assembly forms a center of gravity in the manner explained so that the intensity of GA of a galaxy or of the universe, while beyond comprehension, are established in the same manner. Also, while the force of gravity has the potential to do work as kinetic energy, it is the energy inherent in elementary particles that is expended to establish the kinetic-energy potential of the force of gravity.

It is the force of gravity acting to bind the neutrino-size elementary particles that results in the creation of neutrons and eventually their transformation into protons. (In the process assemble to form neutrons and consequently protons, the elementary particles align in a manner defined as quarks. That process is more definitively explained by the blog entitled Explaining Quark Behavior.) Elementary particles are the instruments by which the fundamental energy is carried to protons that is then transformed to produce photons. A fuller explanation of protons and their behavior is provided in a subsequent chapter.

While photons have absolute behavior characteristics and are assembled attraction energy charges of the fundamental energy that powers elementary particles, the charge intensity of photons is not a precise set value. Photons are assembled by protons or by the proton center of atoms and molecules where the charge intensity of the involved proton center is a function of the number of elementary particles employed in making the atom or molecule. That is why the energy intensity of each type atom or molecule is unique and why the charge intensity of photons can be used to identify the type atom or molecule it came from. Photons are assembled by protons. The power intensity of a proton center is reduced by the amount of heat energy taken up and by the inverse squared rule. The resulting power intensity of a particular proton or proton center establishes the power intensity of the photons produced by a photon center. Not only does unique photon power intensity establish the binding force produced by each unique situation, photon power intensity exhibits a unique measurable frequency or wavelength (explained more fully in a subsequent chapter). The frequency or wavelength of streaming photons is an indicator of photon intensity. It is the wavelength set by streaming photons that produces a color unique to each frequency.

The absolute character and behavior of the described elemental matter and energy forms the basis for the physics principles inherent in creation. Nothing described should be blasphemous or otherwise contrary to fundamental religious or physics beliefs. Both derive their origin in one common event—the creation of what exists. Neither existed prior to that event. Physics is the study of the creation processes. Creation was not a religious event. The concept of God and creation as a religious event developed following the creation of human beings and exists as a faith-based belief. Belief in a creator God does no harm and has the potential to produce much good. Creation is real and exists whether or not the concept that the creator is God is believed or not. To argue otherwise unnecessarily generates divisiveness that adversely affects understanding of both religious and physics principles.

Understanding the fundamentals of the physics involved in creation makes both religious mysteries and mysterious physics principles more readily understood, in fact, enjoyably meaningful. Even if not interested in or capable of understanding the mechanics and mystery-clearing power of fundamental creation physics, just to recognize that the processes of creation are not mysteries but are explainable in common language satisfies a yearning; it is rewarding.

CHAPTER TWO

Presence Energy

The question of whether or not there is a source of energy exists to power creation and the sustenance of creation is at the heart of the anticreation bias that exists in physics. That question was argued from the earliest days of the study of physics by those unable or unwilling to accept the biblical claim that a creator powered creation. After generations of arguing, the conservation of energy theory was adopted as a law of physics following acceptance of the kinetic theory of matter developed by Bernoulli in 1738 and publication of the conservation of energy theory in a book in 1847. The law of conservation of energy says essentially that all the energy needed to create and sustain the universe comes from the matter of which the universe is made.

The law of conservation of energy claims that the total energy of an isolated system cannot change, that it cannot be created or destroyed, and that mass and energy cannot be added to and are not removed from the isolated system. It claims that the energy to power a system comes from conversion of the mass of which the system is made, which is a finite quantity. Therefore the system expires when its finite quantity of matter is expended. Upon reflection the conservation of energy law infers that all activity in the universe is self-perpetuating, which is absurd, and it fails to explain where the mass and energy came from to create the universe in the first place nor does it explain the process of conversion.

The unproven concept that mass can be converted to energy and energy to mass without loss is fatally flawed. Research and study has concluded that atoms are the elementary particles of all more complex forms of matter and that atoms are made of elementary particles beginning with those that make protons, neutrons, and electrons. That research also establishes that the assembly of elementary particles to create matter or mass is done using energy powered bonding forces. In other words, mass is not produced by converting mass to produce energy but rather by using the energy that is maintained as the charge on elementary particles to bind those particles to produce mass. In the same manner, mass is not converted to energy. When the force binding assembled particles fails as in combustion or fission the energy that had been producing the binding force is released as photons that radiate from that point to produce light and heat (combustion or explosion). The involved assembly may be dismantled leaving less complex assemblies or leaving unbound elementary particles but mass is matter is not converted to energy. The elementary particles that form mass remain available to be reassembled, they are not converted to energy. Since elementary particles that are mass did not convert to energy to produce binding force and the energy of the failed bonds is released and radiates to produce light and heat so is no longer available for use in establishing new bonds, the conservation of energy law is flawed.

Theoretically the energy to create and sustain creation exists because the universe exists and is waiting for proof to be developed and the source recognized. The fundamental energy of creation, the energy that powers gravitational attraction, is not electrical energy and cannot be detected using instruments and technology designed to detect electrical energy. However, proof is evident by recognizing that the force involved in producing gravity and the binding of elementary particles exists. What is being offered as an explanation for the source of the energy of which elementary particles are made is not and possibly may never be proven with empirical evidence. However, there exists accepted theory to substantiate the concept. Consider that a space the size of the universe completely void of anything is filled with absolutely identical particles. Those particles distributed ubiquitously throughout the universe space constitute an enormous presence and equally enormous energy potential and would continue that potential as long as that presence existed. The difference of potential between adjacent particles would be exceedingly small but across a vast expanse would be significant. That potential would exist as a charge field not unlike the Cosmic Background Radiation charge field. Such a charge would exist ubiquitously with the presence particles providing instantaneous energy. The term *presence energy* is used by the author when referring to that form of energy.

Presence energy is not radiated. It is sustained absolute by each elementary particle, and because radiation is not involved, it is instant-acting. In an assembly of elementary particles, the power intensity of the attraction component of presence energy is additive in proportion to the number of elementary particles assembled, which collective energy charge would act from the center of the assembly. Without a repelling component, the attraction component of presence energy forms a singularity at the center of the assembly, a center of gravity. As a consequence, each *assembly* of the elementary particles acquires attraction energy intensity equal to the sum of elementary particles making that assembly. Presence energy that exists as a component inherent in each elementary particle is radiated from its center of gravity, a singularity where the charge relationship between particles always exists and is instantly and continuously available. Even though acting instantaneously and in a straight line, the charge field forms a sphere that surrounds the assembly where the intensity of the charge diminishes as the square of the distance from the center point from which acting. That behavior is known in physics as the inverse distance-squared rule. The trinity of charges carried as elementary particles attracts or repels other like elementary particles only, which explains the findings of Galileo at the Leaning Tower of Pisa.

The power of an elementary particle to attract another elementary particle is established and sustained by a *charge* that, in addition to empowering the ability to attract, establishes on each elementary particle the power to repel like-charged particles. Then through consideration for the behavior of behavior of elementary particles the conclusion is driven that a third energy component. The third charge establishes an affinity of the attraction charge and the repelling charge, which causes the attraction charge component to be surrounded, encased by, the repelling charge component. The repelling component of the *presence charge* opposes the attraction force, which prevents the collocation of elementary particles and by that action introduces distance and time as functions in physics. Because the intensity of the attraction charge diminishes with distance from the center of gravity, the power available to produce force decreases with distance, which causes the inverse distance-squared rule phenomenon, and it introduces time as a factor in physics. While it is the opposing attraction and repulsion forces, the yin-yang forces that establish *physical presence* it is the affinity charge attracting between the attracting and the repelling charges that encapsulates those forces to create photons and lines of magnetic force. When assembled by the trinity of forces the resulting particle becomes matter, which translates to mass.

Each elementary particle added to an assembly contributes one unit of attraction power intensity to that center of gravity. With one unit of attraction power for each elementary particle, the attracting power intensity of the center of gravity of an assembly of elementary particles is a reflection of the number of particles and thereby the mass of the assembly. Elementary particles are the carriers of the trinity of components of presence energy that is transmitted to any subsequent assembly of elementary particles. In other words, presence energy is a component of each elementary particle, and that explains the belief expressed by the kinetic theory of matter that the energy to power the behavior of matter is inherent in the elementary particles of which it is made. From that aspect, the kinetic theory of matter is correct. Interpretation and explanation of the processes and cause are flawed.

Elementary particles self-assemble through the action of their attracting and repelling forces wherein the addition of each elementary particle to the assembly adds attraction intensity to its center of gravity of the assembly. Each additional presence particle also adds to the physical dimension and mass of an assembly. As a consequence of the inverse squared rule, progressive increase in physical size coupled with a decrease in power automatically limits the number of elementary particles that can self-assemble. Research has established that the maximum size for self-assembled elementary particles is 1,839 (there is some disagreement on the exact number), which size assembly has been designated a neutron. As a consequence of assembly, the neutron acquires a center of gravity, which has an intensity of 1,839 units of attracting power or a mass of 1,839 units. Current practice uses the term *atomic weight* to denote mass.

At this point, it is necessary to digress to clarify the term *mass*, which by current practice has conflicting interpretation. As used herein, mass refers to the *quantity* of matter, the number of elementary particles in an assembly that current theory call rest mass. That definition translates to mean *the intensity of attraction power of the center of gravity of an assembly*, whether resting or traveling at the speed of light. However, because the intensity of attracting power is diminished as distance from the center increases, the attraction force produced by an assembly does not translate into actual mass or the power intensity of the attracting force acting on one assembly by another except and unless the distance factor is taken into consideration. That factor was not taken into consideration when the periodic table of the elements was created, which means another reason the periodic table of the elements is flawed.

Einstein's theory of general relativity interprets the term *mass* to mean the cumulative force of gravitational attraction and acceleration even though force produced by acceleration is not force produced by gravitational attraction (mass). Acceleration is a consequence of *mass in motion* with motion that may be produced by the expenditure of energy other than the energy that produces gravity. Mass as defined herein is a consequence of a fundamental absolute, so it is an absolute itself and must stand independent of other forces that may be acting on an object if the term is to have any validity.

Now back on the subject. Neutrons are not stand-alone stable. Without outside interference, the self-attracting force of the elementary particles assembled to make a neutron assembly will eventually overpower their self-repelling force, and the neutron transforms into a *proton*. The process of neutron transformation is explained in a subsequent chapter. Protons are the principal building blocks of atoms. Atoms are an assembly of various numbers of protons and neutrons bound by gravitational attraction and the lines of force produced by protons.

In the process of neutron assembly the elementary particles assemble in response to the trinity of energy charges where the intensity of the center of gravity formed by attraction charges increases as the assembly grows in number and physical dimension. A point is reached where because of the inverse distance squared rule, the assembly lacks the power to grow. Evidence suggests the assembly at that point is what has come to be known as a *quark*, which is 1/3 the size of a neutron but with an attraction intensity of 613 units. With that intensity the first quark aids in the assembly of a second quark that grows to its 613 unit limit. Then a third quark is formed where growth is stopped because the physical size of the 3 quark assembly is limited by the inverse distance squared rule. The quarks are formations that exist only within neutron and proton assemblies (all hadrons). They are not real standalone particles but because of their formation establish three X613 unit attraction power centers that multiply the attraction power of neutrons and protons over the power of an assembly of 1,839 elementary particles. Because of the quark phenomenon the power intensity of a center of proton power is a function of the number of assembled elementary particles plus the intensity increase acquired by the quark formation phenomenon. Furthermore, the power intensity of assembled protons is additive, and the center of proton power is in addition to and acts independent of the center of gravity of an assembly. Also, an assembly of protons, atoms, and molecules acquires as many lines of force as there are protons in the assembly, but the attracting energy intensity of the photons produced by assembled protons varies. Photon intensity is a function of the number of assembled protons, the reduction in power due to distance over which each line of force is operating, and the loss of power intensity due to heat. (The loss of power intensity due to heat is covered in a subsequent chapter.)

There is another behavior that influences the functioning of protons. The innermost orbit of an atom would be the smallest diameter and therefore is near full intensity with the intensity less with each additional proton because of the inverse squared rule. (It should be noted that with a helium atom with two protons at equal but near zero diameter orbit distance, the cumulative intensity is applied near full strength to both protons. As a consequence, helium atoms are tightly bound and are the foundation on which all larger and more complex atoms are built. Also, helium nuclei stripped of electron-binding power, as occurs in nuclear or cosmic reactions, are involved in producing destructive beta particle radiation.

The intensity of the attracting power of photons produced by protons varies relative to the structure of the atoms and molecules that produce the photon. As a consequence, while all are presence energy photons, they exist in different intensities depending upon their origin. In fact, the attraction energy intensity of the photons in each orbiting line of force of each type of atom or molecule is unique to that atom or molecule, and each varies further depending upon its prevailing temperature. The wavelength of radiating photons, which is explained in a subsequent chapter, is a distinct indicator of the intensity of the energy carried by a stream of radiating photons.

Presence Energy Summary

- As a component of elementary particles, the attraction component of presence energy is an absolute sustained finite quantum of energy that empowers elementary particles with mutual attraction.
- Elementary particles are the carriers of presence energy that protons transform into photons.
- Photons are the carriers of presence energy that provide the constricting or binding force in the electromagnetic lines of force produced by protons. These are the lines of force that bind elementary particles to make atoms, atoms to make molecules, and molecules to create all things including all higher-order forms of energy.
- Streams of photons created from presence energy recovered from the elementary particles of which protons are made are encased in the repelling force charge (electron charge), also recovered from its fundamental particles in the course of being transformed and radiated by protons to form electromagnetic lines of force.
- Photons are released to radiate when electromagnetic lines of force binding atoms or molecules are broken as in the flame produced by combustion or the flow of electron charges through a lamp filament. Radiating photons, now called light, are not visible so should more appropriately be called illumination or illuminators.
- When radiating photons impact pigmented atoms or molecules, their radiation is stopped, and the energy they carry is released. The energy being presence energy is visible and, as a consequence, make

the matter upon which they are deposited visible as well. That is illumination, the effect called light.

- Attraction energy, of which photons are made, when released onto matter to produce illumination, loses its power of attraction.

- After losing the power to attract, the residue of the attraction component is taken up by proton centers whereby the attracting power of that proton center is reduced and weakened by a comparable amount. That is the process that produces the effect called heat, and the potential to produce the effect called heat is called temperature.

- Expended attraction presence energy or now heat-producing energy, can be and is transferable to proton centers exhibiting stronger and more intense proton center power. Since proton centers that exhibit greater intensity attract the residue that produces heat, they are deemed to be colder. That is, the phenomena recognized as heat flows from hot to cold.

- Expended presence energy, now heat-producing energy, can be contained by a magnetic field (covered in detail in a subsequent chapter), as photons are contained by the magnetic field that encompasses the electromagnetic lines of force produced by protons.

- The lines of force produced by a proton whose proton center is weakened are then more easily broken, and if the expended presence energy (heat) is of sufficient quantity, it will cause other bond failure. Failed bonds release their photons to produce more heat as happens in combustion where the possibility exists for a chain reaction as long as the supply of vulnerable matter exists.

- Overheated proton centers lose the ability to produce photons and, as a consequence, lines of magnetic force and, in turn, the ability to form atoms and molecules or to produce heat. Matter in that state would exist bound by their center of gravity into a compact, very dense assembly of bare dormant protons to form the core of large bodies such as stars (especially dead stars) and galaxies.

- The belief that all matter is made of presence energy particles and that presence energy is present in all things is a belief common to both religious and science. The initial effort in Genesis to explain creation and Abraham's designation of the almighty presence as God was published, along with other beliefs and experiences, in what became known as the Bible. Refusal of the community of scientists to accept the Genesis version of creation resulted in the development of a conspiracy to disavow the existence of a source of energy that powers creation.

CHAPTER THREE

Proton Behavior

A visual illustration of proton behavior does not exist and would be meaningless without a narrative explanation since the components are not visible and have no electrical charge so are not detectable using electrical charge detection equipment. The elemental components of neutrons and protons are detectable only because each elementary particle of which they are made constitutes one unit of gravity attraction or mass. This narrative explanation has been developed through application of old-time logic or a priori consideration of the principles and processes that would be needed to produce the behavior associated with protons.

The behavior and role of protons is extremely complex and absolutely vital to the formation of matter. Protons are the heart of creation and the building blocks of all matter and all forms of energy beyond the elemental form, gravity. Fundamental physics, and therefore all physics, cannot be understood without first understanding how protons function and what their role is. The following is a narrative picture developed in an effort to explain proton behavior.

Begin by visualizing a universe filled with nothing but invisible elementary particles (neutrinos) dispersed and held in suspension by their interacting attraction and repulsion (yin-yang) forces, in which this interaction is powered by a ubiquitous eternally maintained presence charge (cosmic background radiation?). That is the first step in the creation of physical matter (mass). Then visualize that the attraction force between two of those particles overcame the repelling force so that those two particles were bonded into an assembly by their inherent affinity for each other, in which this assembly then has an attraction power intensity and a physical presence two times that of a single neutrino but with no increase in repelling power.

An assembly with a higher intensity center of gravity begins to progressively grow. The attraction intensity of the assembly increases with each added neutrino until the distance from the assembly center to its periphery reduces the power of attraction (the inverse distance-squared rule) to the point where insufficient power exists to bind any additional neutrinos into the assembly. With radiated energy, the attraction power intensity at a distance from the source is inversely proportional to the distance squared. The assembly acquires physical presence and dimension because of the interaction of the attracting and repelling (yin-yang) forces. Research has established that only 1,840 (some disagreement on the exact number, but the number is absolute) neutrinos can assemble in the process described, and the assembly at that point was given the name neutron. A neutron has atomic weight (mass) but is otherwise neutral because all attraction power is expended in its assembly and a proton center has not been established. With 1,840 neutrinos, each neutron has 1,840 units of mass (attraction intensity) at its center of gravity. That is analogous to the atomic weight measurement used in the periodic table of the elements. Recognition is given to the phenomenon whereby elementary particles assemble as a trio of intermediary formations called quarks.

Neutrons are an intermediary step in the formation of protons. Neutrons are not a stable assembly because the interaction of attraction-repulsion forces within the assembly is a continuing battle. As contraction (compaction) progresses, the distance between the center of gravity and the fundamental particles decreases, and the applied attraction force strength increases. Consequently, the power of attraction grows to the point where eventually the interparticle repelling force is overcome. The repelling charge of the last neutrino is ejected from the assembly, allowing the center of gravity to succeed in pulling the *affinity charge* off the 1,840 neutrinos. Freed of the *repelling charge* component, the affinity charges form a singularity with an intensity of 1,840 units with the 1,839 repelling charges surrounding the attraction charge singularity. Considering the quark phenomenon, the elementary particles for three intermediary assemblies that multiplies the force producing power of the proton assembly.

The term singularity is used because with no interacting repelling force, the involved energy charges exist at a single point with no intervening space or distance, where interaction is instantaneous, and the intensity is sustained as constant by the existence of the elementary particles. This singularity exists in addition to the assembly center of gravity. The transformed neutron is called a proton, and the singularity of attraction charge gives protons the power to radiate photons (attraction energy in quantum packets), each with 1,840 times the power intensity of a single neutrino. (The quark phenomenon may multiply the attraction energy component.) The transformed energy radiated by protons formed into lines of magnetic force that not only function to bind the assembly of atoms and assemblies made of atoms but the other components of which made, the repelling anti-affinity electron charge and the attraction or affinity photon charges, provide the means of producing the effects called electricity, heat, and light. The lines of magnetic force emitted by protons cause what is called magnetic force or magnetism. The lines of force emitted by protons are the conveyers of the energy that is acquired from the elementary presence particles (neutrinos) and transformed into photons. The energy carried by lines of magnetic force powers all subsequent forces and effects, including the *giving and sustaining of life*. Protons are the elementary particle, the building blocks of atoms.

As an aside, those familiar with the principle involved in electric motors may have noticed that the 1,840 affinity charges of the proton singularity and the surrounding 1,839 anti-affinity charges of the proton core have created in effect an electric motor where the surrounding anti-affinity (repelling) charges rotate (spiral) around the singularity. Note then that the surrounding anti-affinity charges prevent the singularity from radiating *except* along the axis formed by the spiraling anti-affinity charges. For that reason, proton radiation is a single line of affinity energy particles (photons), each particle with 1,840 units of intensity. In the process, the line of radiating photons is encased in the anti-affinity or repelling charges (electrons) that are supplied by the charge center that surrounds the singularity. As a consequence, the anti-affinity encased line of radiated photons is attracted to the opposite pole of the involved proton, forming a loop as seen in illustrations of lines of magnetic force. The spiraling looped line of anti-affinity encased photons radiated by protons produces two forces. The encased photons (affinity charge) that make up the core of each line of force attract each other to produce a constricting (binding) force. The spiraling anti-affinity outer cover establishes a repelling force between like lines of magnetic force so that no two lines can occupy the same space. The repelling energy charges confine the attracting charge particles, photons.

As an additional behavior, the repelling (electric) charges that encase each line of magnetic force are the mechanisms involved in producing electricity. In the generation of electricity, the repelling charges (electrons) are transferred from lines of magnetic force to an isolated insulated conductor in the generator and the affinity charge is set up to drive the electrons from atom to atom as they are pushed by the affinity charges. Before passing on to other subjects, recognize that the energy contributed by the prime mover of the generator is expended in the work involved in transferring the electric charge from lines of magnetic force to the generator conductors and in overcoming the resistance to the movement of electron charges over the electrical system conductors. The electron charge energy that does the attracting and repelling work electrons do is presence energy supplied to and through the elementary particles.

(The implication of spiraling lines of magnetic force has not been addressed, but it is postulated that behavior seen in the earth's magnetic force field is involved in the formation of tornados, hurricanes, and other weather phenomena including low-pressure cells.)

Recognition of a relationship between protons, electrons, magnetism, light, and heat is not new, but that relationship has never been clearly explained because it has never before been understood. The concept that protons produce lines of magnetic force developed after the process of creating protons from presence energy charged elementary particles was recognized through a priori reasoning. A lifetime involved in generating electricity made quite clear that electricity, heat, light, and magnetism are products of a common phenomenon. From there, old-time logic led to the realization that the commonality of those forms of energy exists in lines of magnetic force (LMF) assembled from the trinity energy components. Presence energy is the energy carried by the elementary particles of which protons are made.

It is the constricting power of LMF by which matter is assembled. It is the repelling charge that encompasses the streaming photons emitted by protons that produces the power attributed to electrons and electricity, the charge stripped off LMF in a generator. It is the affinity charge transferred to generator conductors along with the electron charges that creates the difference of potential that drives the flow of electricity. It is the attraction charge of fundamental energy assembled into photons that produces the effects called light, heat (as explained in a subsequent chapter), and lightning, which is not electron charge energy but the attraction charge component of presence energy or heat energy.

The significance of that realization is set in perspective with recognition that there is a lot of heat-producing energy in the world that could be harnessed with an appropriate system to produce photonicity (a term coined for this purpose) with potential paralleling or perhaps exceeding electricity. Understanding the energy that produces heat along with recognition that heat energy can be contained and managed using magnetic force fields (a fuller explanation in a subsequent chapter) should lead to the development of photonicity as a viable alternate energy. This is knowledge hidden by the anticreation conspiracy put in place with the conservation of energy law to disparage the concept of an almighty creation powering energy source.

CHAPTER FOUR

Defining Heat

With all the technological advancements in disciplines where heat is a factor, it is rather difficult to accept that the fundamental physics involved with heat could be so poorly understood, but that is precisely the objective of this argument. Clarification begins with recognition that the effect called heat is a process that follows the illumination of matter with the energy carried by and only by photons. This is a complete departure from the current theories espoused by the community of physicists and the academia that teach the current theories exclusively throughout academia.

Modern College Physics, a college-level textbook published by D. Van Nostrand Company Inc. and authored by Harvey White, professor of physics of the University of California, explains heat as, "According to the Kinetic Theory of Matter the individual atoms of which all substances are made are in a state of rapid motion. As a body is heated to a higher temperature this atomic motion increases and the body expands. As a body cools the atomic motions decrease and the body shrinks. That heat is a form of energy and is due to the kinetic energy of molecular motion first proposed by Benjamin Thompson in the later part of the 18th century." *Merriam-Webster* defines *heat* as "to become hot or to make hot." *Webster's New College Dictionary* says "the quality of becoming hot." Eventually *Wikipedia* explains, "In Physics and Chemistry, heat is energy transferred from one body to another by thermal interaction." The current theories, understanding, and definition for heat and the process involved stand with those archaic pre-atomic age theories that were accepted without empirical scientific evidence and subsequently drove the development of the flawed kinetic theory of matter and the law of conservation of energy.

The modern atom theory of matter brought realization that the elementary elements that all matter is made of are themselves made of subelements and that the elements and subelements are bound into and held as assemblies with force produced by particles (quanta) of energy. Unfortunately, the requisite for energy-driven binding force is generally not recognized as a constituent component of matter/mass with the archaic kinetic theory of matter and the law of conservation of energy considered as obligatory.

For some reason, possibly driven by the anticreation mind-set that permeates science beliefs, reevaluation of the theories regarding heat in consideration of the atomic theory of matter has been overlooked.

An enormous amount of research and observations of processes related to the atomic theory has been compiled. Reassessment of the processes revealed leads to the conclusions expressed here.

> First and foremost, supported by a myriad of empirical evidence, the energy that establishes and sustains the bonding force involved in the assembly of matter (mass) is a constituent part of the assembly of matter or mass every bit as much as are the protons, neutrons, electrons, and atoms.

> Second, the evidence collected shows that when the bonds involved in binding atoms and assemblies of atoms are fragmented for whatever reason (whether combustion, physical force, effect of heat, electron flow from atom to atom along a conductor especially a filament, photosynthesis, sunlight, or nuclear fission), photons are released to radiate at the speed of light from the point of bond failure. That leads to conclusion 2a, photons are a constituent part of the lines of force that bind matter/mass, and conclusion 2b, and photons are the energy that powers binding.

> Third, a myriad of empirical evidence establishes that when radiating photons impact a pigmented atom or molecule, the energy quanta of which photons are made washes over the material upon which they are impacted, making the impacted material visible. The degree of illumination is a function of the number of photons impacting and the intensity of the energy quanta carried.

> Fourth, empirical evidence shows that following illumination, the energy that produces illumination is taken up by the center of proton power of the involved atom or molecule, which precipitates

the effect called heat. With reasoning, the evidence leads to the conclusion that the ability of a proton power center to produce binding power is reduced when the energy residue of illumination is taken up to the point that the bonds formed by that proton power center are weakened, the bound matter expands, and when the amount of heat energy is of sufficient quantity, the bonds are fragmented to release more photons and produce more light and heat. This process is seen in all actions involving the fracture of atomic bonding.

Cursory analysis brings realization that the energy that produces heat is released from the lines of force produced by protons that bind atoms and assemblies of atoms. The lines of force produced by protons that are packets of the attraction component of presence energy wrapped in the repelling component of presence energy, are frangible. When lines of force that bind inter-atom bonds fail, the contained photons are released to radiate from the point of failure at the speed of light so are not involved in establishing new bonds nor are they in the form needed to create new bonds. This is best illustrated by photons that are released when an electric current is caused to pass over an incandescent lamp filament. As electron charges flow along the filament, electron bonds are broken and new bonds made as the electron charges are forced to move along the filament from atom to atom. The lamp filament remains intact while atom to electron bonds are broken, which means the photons that produce illumination are released from the failed bonds. Each broken bond releases the energy that powered the bond in the form of photons that radiate from that point at the speed of light. For each ampere of current flowing through the filament of a light bulb (about 120 watts for a 120-volt bulb) there are 6.24×10^{18} electron charges passing a given point each second, so there are that number of bonds broken and new bonds made every second for each atom along the length of the filament. The energy lost to radiation is not available to form new bonds nor could it power the formation of new bonds because the released energy is no longer in the form or the position to establish bonds. New bonds are formed by new lines of force emitted by the proton power center of the receiving atoms.

Following bond failure, the released photons radiate at the speed of light until they impact pigmented atoms. That can be atoms of molecules in the atmosphere or atoms making up larger objects. Upon impact, the energy the photons are made of illuminates the matter upon which it is impacted. When the number and intensity of the photons is sufficient, the photon energy is visible and therefore makes the object visible. Whether producing visibility or not, illumination is a transitory process where the object remains illuminated only so long as the flow of photons continues. The energy released when photons impact pigmented molecules or large objects is soon taken up by the objects' proton power center to produce the effect called heat. That is why sunlight produces heat. The point that the effect called heat is produced by the energy radiated as photons. It is not heat that is imparted by the photons; it is the energy that produces the effect called heat, and the heat effect is not produced by the rapid motion of atoms. The process described holds for all heating wherever radiating photons impact on pigmented matter whether or not of sufficient magnitude to be visible or at a point not visible, such as inside a cell inside our body or because the massive quantity of bond failure hides the process such as in a nuclear reaction or the sun.

Heat is a factor in every aspect of creation powered by proton energy and plays a significant role in many phenomena besides combustion and combustion-related processes. Gravity and the energy that powers gravity are not involved in producing light and heat but are the source of energy that powers protons that are involved. Heat and the energy that produces heat are major players in weather phenomena. It is the heating process described that produces violent weather such as thunderstorms, hurricanes, and tornados. It is the energy that produces the effect called heat that is seen as lightning, and whether in the most complex weather systems or the most complex solar system, the fundamental principles of the heating process are absolute.

It is difficult to comprehend that the most brilliant minds are unable to recognize the processes involved with the production of heat. However, the processes outlined are not revealed without it being first recognized that the processes of light and heat are powered by a three-component energy charge that is implanted and sustained on the elementary particles of which all things are made. It is considered unlikely that the anticreation mind-set exists so consistently across the community of scientists solely on the basis of religious beliefs. That leads to the realization that the fundamental physics principles involved in producing heat are hidden from even great minds who do believe in creation by the anticreation conspiracy inherent in the kinetic theory of matter and the law of conservation of energy.

Following acknowledgment that the behavior of matter is powered by energy implanted on and made an inherent property of each elementary particle of which matter is made comes realization that the kinetic theory of matter and the law of conservation of energy are defective and responsible for the development and acceptance of defective theories, most notably those that attempt to explain gravity, light, and heat, which are matters fundamental to physics.

Foremost, with the horse put in front of the cart, it will be recognized that heat is *not a form of energy* but an effect produced by particles of presence energy released from photons that impact matter to produce illumination. The energy expended in illumination is taken up by and thereby weakens the power of proton power centers that, in turn, weaken or cause the failure of force lines that bond atoms and assemblies of atoms. When force lines weaken, atoms expand, and when the bonds fracture, the elementary particles are released from that bond; the assembly fails with that process called combustion or fission. Additional heat-producing energy is released, producing further expansion of matter, especially the atmosphere immediately surrounding the point of release. The force of expansion that is a result of failed atom bonding is a force capable of doing work, but that force is not a form of energy. The heat effect is produced because of the photon residue taken up by proton energy centers.

The unproven theory that heat is a consequence of atoms and the presumed particles of which they are made being in constant motion that increases with the addition of heat and diminishes when heat is removed is wanting in many ways. That explanation fails to establish what heat is, how motion causes heat, what powers the constant motion, and if produced by the collision of particles in motion, how. What is the process? And most of all, what powers the constant motion? It is mind-boggling trying to understand how the most brilliant minds can accept that explanation, especially considering how heat is such a significant factor in physics. It goes to show the mind-consuming power of the anticreation conspiracy inherent in physics.

The concept of presence energy recognizes that supplying the energy fundamental to creation is a continuing, endless process. That may account for the conclusion that atoms were thought to be in a continuing state of motion as expressed by the kinetic theory of matter, but there is no evidence that movement of energy produces heat or the continuous physical motion of atoms. It is conceded that variations in the intensity of the energy that powers bonding could and would cause the involved articles of matter to move as occurs when proton power centers take up the energy residue of illumination.

In the process of transforming neutrons into protons, protons acquire behavioral characteristics that are absolute and unique. Protons assemble the attraction component of the presence energy carried by elementary particles into photons that, in the course of being emitted by protons, are assembled into streaming (constant state of motion) lines of magnetic force (LMF). The attracting energy charge is formed into photon particles possessing the attraction intensity of 1,840 gravitons plus three 613 unit quarks. In the process of emission by protons, the stream of photons is surrounded (encased) by the repelling electron-charge component of the fundamental energy charge, carried by elementary particles, to form LMF. While LMF are the medium by which all matter is bound, a process fundamental to creation, LMF are frangible. The release of the energy remaining following illumination is the number one cause of the failure of inter-atomic bonds although physical force, such as that produced by gravity, causes inter atomic bonds to be fragmented.

Photons are released when LMF are fragmented, and they radiate to produce the effect called light, which is covered in the subsequent chapter. LMF also power reactions referred to variably as heat, chemical, metabolic, and meteorological reactions; combustion; fission; and nuclear and atomic energy. Each of those reactions involves the breaking or weakening of the bonds that hold together the atoms and molecules of which LMF are a constituent part. Bonding force lines are weakened to the point where they fragment when the contributing protons are heated as occurs in combustion. LMF are also broken to release photons and produce light and heat with the passage of electric current along specially selected conductor materials such as lamp filaments and electric heating elements.

The process involved is easily understood as long as the horse is kept in front of the cart. Radiating photons pass through transparent material without impact, but photon radiation is blocked upon impact with pigmented atoms and molecules. Upon impact with pigmented matter, the graviton energy charges that make up photons, the attraction component of presence energy, are released onto the impacted material. The energy of which photons are made is visible when sufficiently concentrated and so illuminates the matter upon which it is released, making that matter visible. Second, subsequent to producing illumination, the remnant energy that gave up its mutual attraction power to power illumination is taken up by the protons or proton center of the atoms and molecules upon which impacted. The acquiring proton center expends a commensurate portion of its attraction energy potential to attract and bind the expended graviton energy. In other words, the intensity of the acquiring proton center is reduced by an amount commensurate to the amount of energy taken up, thereby weakening the bonds being maintained by that attraction center. The weakened bonds result in expansion of the involved atom or molecule. *Expansion* is the first indicator of the effect called heat. Heat is not the product of the nebulous theory of molecule vibration or atoms in motion. Heat is the effect produced when the attraction component of presence energy is released to illuminate (whether or not visible) and is then taken up by the proton center of atoms and molecules. Heat is the effect caused by the weakening and breaking of the bonds that bind atoms and assemblies of atoms. Heat causes the fragmenting of LMF wherein the energy that produces heat is released, a phenomenon that results in the reactions of combustion and explosion.

When weakened sufficiently, the LMF bonds of the involved atoms or molecules are fragmented, releasing photon radiation, producing illumination plus a heat-induced effect that when sustained is called combustion or explosion. The reaction is the same whether called chemical, combustion, fission, nuclear, atomic, or plasma. It is the same whether occurring in a candle, an atom, or a sun. Bear in mind that the intensity of the energy contained by a photon can vary significantly depending upon the force the LMF of a component was applying at the time the line of force was fractured. For example, the force applied in bonding hydrocarbon fuels is considerably less intense than is the force applied in bonding uranium atoms or atom nuclei. While differing in intensity, proton energy is in every instance the same attraction component of presence energy extracted by protons from the elementary particles of which the protons are made, whether in the flame of a candle, an atomic explosion, or a sun.

Heat-producing energy is attracted to and taken up by the atom or molecule with the highest exposed proton-attracting intensity. That behavior is the reason heat energy is transferred from hot to cold. Modern physics identifies matter projecting *higher* attraction intensity as being colder and matter projecting *lower* attraction intensity as being hotter. There is a correlation since the more intense proton power centers have greater heat energy attraction, translating to less retained heat energy, and less powerful heat-attracting centers have a greater retained heat energy. The magnitude of heat energy retained by an assembly of atoms, or matter, establishes the ability of an assembly of matter to cause heating through convection to other matter. The extent of that ability is called temperature. For example, a thermometer measures the temperature of a body as 98.6°F because the thermometer is calibrated to translate the degree mercury expands at that temperature. Mercury expanded because it took up heat transferred from a 98.6°F body. Maybe that is why temperature is measured in degrees.

It goes without saying that the development of technology relative to controlling and utilizing heat is adversely influenced by lack of complete understanding of the processes involved, and that lack of understanding is the direct result of the anticreation conspiracy imposed by modern physicists. Failure to understand the processes involved with heat explains why technology to utilize directly the energy that produces heat and light has not been developed. Failure to understand the processes involved in producing heat also limits the ability to understand the burning bush seen by Moses and the flaming furnace of Meshach, Shadrach, and Abednego.

CHAPTER FIVE

Global Warming

Ever since the hydrogen bomb proved successful, there has been research ongoing at many locations around the world all trying to find a way to control nuclear fusion in the expectation that energy would be released for use commercially as is the energy from fission (atomic) reactors. After forty-plus years of intensive research employing the best minds in the world, those efforts have gone without success for a reason. Energy is not released by fusion reaction. However, in the course of that research and experimentation, a reaction-containment system was developed using magnetic force fields that have proven capable of containing the heat (150 million degrees Celsius) and pressure involved. It is the tokamak reactor developed by ITER (International Thermonuclear Experimental Reactor) at a location in France. The success of the tokamak reactor opens the door to a new insight on global warming.

New windows of understanding of global warming open with realization that heat is an effect caused by the behavior of a form of presence energy that is carried by photons. Photons are released to radiation when the lines of force that bind the components fundamental to all things physical are ruptured as occurs, for example, in the process of combustion or in nuclear fission reactions such as occurs on the sun and nuclear reactors. When the understanding revealed by the tokamak reactor's success in containing high-temperature plasmas is applied to analyzing global warming, never-before-appreciated interaction between heat and magnetic force fields is revealed. The tokamak reactor project has demonstrated that a magnetic force field is capable of and can contain heat.

41

With better understanding of the processes and the behavior of heat, very significant knowledge is exposed that provides a more meaningful understanding of the cause of global warming and gives promise for the development of revolutionary new methods to manage and control heat. Heat energy exists in abundance everywhere on earth. It is visible. It is sunlight, lightning, lava, and it causes the wind to blow. The technology is begging for development. All that is missing is the understanding that is being blocked by the anticreation conspiracy put into place by the very people that profess their mission is to develop understanding and education.

The lack of understanding of the processes involving heat is driving the efforts to control global warming in directions that are hurting the world's economy and technologic advancement, especially with regard to finding alternative energy sources and ways to mitigate the severity of weather. Severe weather is heat driven and causes loss of life and large-scale economic and property damage. Failure to fully understand the fundamentals of heat has sent the search for alternative energy off chasing fantasies and methods that have little prospect for any significant advancement in technology or contribution to obtaining alternate energy sources.

Combining the understanding of heat fundamentals revealed when recognition is given to the existence of an empowering presence energy with the knowledge revealed by the tokamak magnetic reactor offers great promise for significant breakthrough in these areas. Two problem areas of significant concern immediately come to mind as candidates for the application of that technology. The first is to develop the means to extract heat energy from storm cell systems such as tornados, typhoons, and hurricanes so as to mitigate their severity. Indications are the technology exists that can be applied to achieve that end. Storm mitigation using magnetic force is indicated to be environmentally friendly with little or no adverse side effects. The second area involves global warming.

The phenomenon whereby magnetic force fields contain heat energy has not been considered as a factor in global warming. Two pictures are presented to illustrate how the earth's magnetic force field is a major factor in global warming. One is an artist's depiction of a tokamak reactor that shows how a magnetic force field is used to contain the heat energy involved with nuclear fusion that produces temperatures as high as 150 million degrees Centigrade. A second illustration is an artist's depiction of the earth's magnetic force field. The area of heat containment in each case is highlighted.

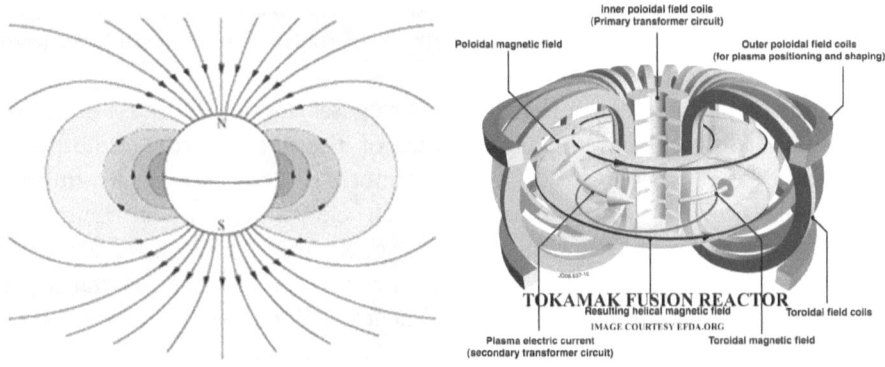

It is important to understand that the heat-containment area is the space within the looped magnetic force lines. On the earth picture, the looped area encompasses the surface of the earth from about twenty degrees north to twenty degrees south latitude. Note that the magnetic pole areas and the magnetic force field axis are not encompassed by the looped force field. In other words, the earth's magnetic force field effectively encases and contains heat to the more heavily inhabited portion of the earth's surface while the magnetic pole regions and the core of the earth are open to outer space. More than that, as explained in greater detail in the chapter on proton behavior, the proton power center at the center of the earth is a powerful attractor of heat energy. It removes heat energy from the magnetic pole regions but is blocked from taking up heat energy from the populated area on the surface of the earth. Recognize that removing heat energy from the magnetic pole areas contributes to reducing the temperature of those areas. Connect that to the fact that in recent years, the north magnetic pole has been shifting from northern Canada toward northern Siberia. That would result in increasing temperature in Canada and the areas such as North America influenced by the temperature of that area. Likewise, the temperature in Siberia and adjacent areas would drop as has been the case since the northern magnetic pole began shifting.

While there is no indication this understanding could produce a way to control global warming, it can prevent the misapplication of time, talent, and capital chasing a solution to control greenhouse gases that is the perceived cause of global warming. In that regard, it should be recognized that greenhouse gases are not the cause of but an effect of global warming. The quantity of greenhouse gases in the atmosphere has increased because the temperature (a reflection of the magnitude of heat energy retained by the molecules that make up the atmosphere) has increased. An increase in heat energy retained by those molecules results in lowering their density that normally would be too high to exist in the atmosphere. By all indications, far more heat energy is injected into the earth's magnetic shield via sunlight than all the man-made sources combined.

There is other evidence that greenhouse gases are not the cause of global warming. Consider that the temperature of the entire earth's atmosphere and surface is raised perceptibly within minutes of being subject to sunlight and cools similarly with sunset. No such change can be attributed to man-made sources. Along with recognizing the true definition of heat, the interrelationship of magnetic force and the heat in sunlight illustrates the direction time, talent, and capital should be applied with regard to finding an alternative fuel (energy) source as well as to help mitigate global warming.

As stated, the energy that produces the effect called heat is the attraction component of the fundamental energy of creation after transformation by the action of protons into photons. Photons, the attraction component of fundamental energy, is involved in powering the binding of atoms and assemblies of atoms and in powering the effects called light (illumination) and heat. By comparison, the repelling component of fundamental energy has been developed to other useful forms like electricity, magnetism, and electromagnetic radiation, and the ratio of repelling energy to attracting energy is 1 to 1,840, meaning the power available as photons is 1,840 times greater than the power available as electrons. Recognize that both have a common origin with similar behavior characteristics. In fact, the attraction and repulsion components interact. The point is that there is an enormous amount of photon-type energy, much of which is wasted in the form of excessive sunlight and resulting in unwanted heat, lightning, lava, and destructive weather. With understanding cleared of the biased anticreation conspiracy, technology can be developed to intercept and convert photon energy into a form similar to electricity to serve utilitarian purposes.

CHAPTER SIX

Weather

The earth's atmosphere is composed principally of hydrogen, oxygen, and helium atoms existing in a vapor state. That is, the components existing as the earth's atmosphere are expanded sufficiently from retained heat energy to produce density within a range needed to exist at prevailing atmospheric pressure. In a vapor state, the molecules of each type of element interact with like molecules only, independent of the molecules of other elements (Dalton's law). It is that phenomenon that explains the behavior of matter in vapor form. It is that phenomenon that establishes the density of the elements that comprise the atmosphere and thereby the atmospheric pressure. Therefore, through the absorption of heat, water molecules whose density is within the range can exist in the atmosphere where water molecules in a vapor state interact to form a cloud even if not visible. In other words, water molecules in the vapor state bond with other water molecules to form an interacting assembly.

The assembly of water molecules becomes visible as a cloud when the individual water molecules are large enough to be visible. That is, the water molecules condense by giving up heat-producing energy. When the binding forces interacting between water molecules are stronger than the force of gravity on individual molecules, the cloud becomes an assembly like a ship floating in the sea. At that stage, the force of gravity interacts primarily with the cloud as an assembly rather than as individual molecules. For that reason, clouds of water vapor can hold enormous quantities of water in vapor form. Water molecules expand as they take up heat energy. As they take up heat energy, their density, and thereby the density of the cloud, is reduced, allowing the cloud to hold more water in vapor form. It is the energy that produces the heat effect that powers the formation of clouds. When the combination of heat and density of water molecules is sufficient and surrounding air temperature is cool enough, storm cell clouds develop. The energy that produces heat is the attraction component of presence energy that is delivered to the water molecules through the elementary particles of which they are made. (The energy that produces electricity is the repelling component [electrons] of presence energy. The energy that drives electrons in the production of electricity is the affinity component.) Both heat and electric energy behave in a similar manner although of different polarity and different intensity. Because the behavior of heat is not understood, the conclusion was taken that lightning is electricity, another case where understanding suffered because of the conspiracy to disavow the concept of creation energy.

Lightning is heat-producing energy that drives the formation of destructive storm cells; knowing this opens the door to developing a method of mitigating the heat energy intensity retained by a cloud system and thereby mitigating the strength of storm cells. One such possibility is shown in blog 3 in part 2.

The assembly of water molecules even in a vapor state develops a center of gravity just as any assembly of molecules. The storm cloud pictured is made of water molecules bound into an isolated assembly by the gravitational attraction center of that assembly. When the self-binding force of the assembly on the individual molecules is greater than the force of the earth's gravity on those same molecules, the assembly becomes isolated, floating in the atmosphere like a ship floating in the sea. It is the cloud as an assembly that is tethered by the earth's gravity rather than each water molecule independently. The density of the cloud as an assembly establishes its position relative to the earth.

As the distance from the earth's center of gravity decreases, the force of gravity increases exponentially. There is a point below which the earth's gravitational force on the individual water molecules of the cloud exceeds the attraction of the cloud assembly center of gravity on individual water molecules. That point marks the base of the isolated cloud assembly. The structure of a cloud assembly relies upon intermolecular binding, and the energy that produces heating is delivered through the fundamental particles that are themselves sustained by the continued supply of presence energy.

A cloud system moving in from a tropical location is losing and using energy as it moves, and it cannot grow unless some considerable additional energy is introduced into that cloud system. It is well established that storm systems grow in intensity and scope when they acquire additional water vapor and heat energy. A current theory is that the additional heat energy needed to grow is acquired from a continuing supply of heat-laden water molecules moving in from a tropical area, but the application of a little old-time logic squashes that theory. As heat-laden clouds move to meet colder cooling air, the heat-laden, low-density vapor molecules condense as heat energy is transferred to the cooler hydrogen, oxygen, and helium atoms. The water molecules grow in density until they reach liquid size and are drawn out of the cloud assembly by gravity as rain. The heat-producing energy transferred to the cooler air molecules is lost from the cloud assembly. The point being made is that when the heat-laden clouds meets cooler air, there is a reduction rather than an increase in heat energy within the cloud system, meaning another source of heat energy is required to sustain and promote growth in the intensity of the heat-producing energy. The addition of more moist tropical air will grow the size of the cloud but not the intensity of the heat-producing energy.

By keeping the horse in front of the cart, the process by which the intensity of storm cells grow is made clear. There is activity within storm cells first in the form of upward and downward airflow produced by differences in temperature. In that process, there are numerable changes in temperature and physical activity so that intermolecular bonds are being broken, which causes the release of binding energy as photons that then impact other molecules to release heat energy that in turn causes more bonds to be broken and more heat energy to be released. When the conditions are favorable, a chain reaction is started that feeds itself and the storm cell with potentially significant amounts of the energy needed to increase storm intensity.

That injection of additional heat energy can be seen when it is recognized that the heat released from the molecular bonds is derived from the source that supplies the energy that creates the bonds, which is the presence energy acquired through the fundamental particles of which the water molecules are made. This is seen as contrary to the *first law of physics*, the law of conservation of energy that sets the foundation upon which the anticreator conspiracy movement grew. The law states that the total energy in an isolated system remains constant over time, meaning energy is conserved, neither created nor destroyed, where the operable condition is in an *isolated system*. In the case explained, the cloud system is not an isolated system when it comes to the law of conservation of energy. The energy that sustains the power of the fundamental particles of the cloud continues to be supplied. If not, the system would fail. Apply some old-time logic, and it is recognized that it is the universe that is an isolated system where the source of presence energy is part of that system. The law of conservation of energy is correct when both factors are recognized. Otherwise, the horse (meaning modern physics) doing the interpretation may have eyes but, with the blinders of the anticreation conspiracy, is not allowed to see.

There is more that can be said on the subject, like funnel clouds form as shown in the accompanying photograph. When the density of a section of a storm cloud becomes so great, the force of gravity pulls that section of cloud down to earth, introducing other interactions between the cloud of charged molecules and the earth's magnetic force field that results in the creation of cyclonic winds. Also, an explanation for thunder may put that subject to rest.

Lightning is the movement en masse of the energy that produces heating from one location to another. The energy that produces the effect called heat is retained by the proton power center of atoms and molecules by the process explained in an earlier chapter. It has been demonstrated and quite well understood that heat energy moves from hot to cold. Retained heat energy is attracted to the proton energy center of the atoms or molecules with the highest attracting power (colder) from the atoms and molecules with the lowest attracting power (hottest). Ergo, a collection of water molecules that are cold, as often exists in thunderhead cells, possesses high heat-energy attraction. That situation describes a difference of energy potential not unlike the difference of energy potential that gives rise to the flow of electron charges. The acquisition of heat energy causes expansion of the acquiring $H2O$ (water) molecules. Acquisition of heat energy en masse produces en masse expansion of the acquiring molecules, and when the en masse acquisition involves movement, then the en masse expansion is a moving reaction. Therefore, lightning involves an en masse movement of en masse physical molecule expansion with a corresponding pressure wave that produces the sound known as thunder.

Cyclonic winds are generated when the funnel cloud of a storm cell is pulled to the earth. That action takes place in the earth's magnetic force field. The water molecules in the cloud are charged with the attraction component and the affinity component of the presence energy supplied through the elementary particles. Charged particles moving through the earth's magnetic force field are made to curve as demonstrated by Dutch physicist Hendrik Lorentz, and that phenomenon is called the Lorentz force law. The force imparting circular motion to the charged particles varies with the intensity of the particle charge, and the force applied to the descending cloud is a function of the number and intensity of the charged particles. The individual particles of the descending cloud are bound by the intermolecular bonds so the force applied to the individual charged particles causes the cloud formation to spiral. Molecules of the other elements of the atmosphere are caught up in the spiraling cloud of water molecules. The spiraling motion of the mass of water and air molecules translates to wind, and that describes a tornado. Hurricanes are an assembly of a number of cyclonic cells that become joined through intercellular gravitational attraction.

If this is not enough eye-opening information, then more examples won't help, but eyes opened with this effort can help open the eyes of others to contribute to unlocking understanding.

CHAPTER SEVEN

Light and Vision

The Presence Energy of Creation

Photo by author.

Just pictures of a simple old-time kerosene lamp whose days are gone and a candle, yet they are showing the presence energy of creation. They provide relevance for the command related in Genesis 1:6, "Let there be light," with evidence that the energy fundamental to all creation is, in fact, the energy that produces light. These simple pictures not only provide visible proof that the presence energy of creation exists, they also open our eyes to see and understand what presence energy is and how it functions. This demonstration shows how the continuing denial of the existence of presence energy by the community of physicists has hidden fundamental facts relating to light and vision as was the case with heat.

The flames of this lamp and candle are produced by what is currently defined as combustion, the burning of hydrocarbon fuels, which means the destruction of bonds between the hydrogen and carbon atoms of which the candle and lantern fuel are made along with the bonds of some of the oxygen of the surrounding air. The energy that powers the bonding of the elementary particles of which atoms are made and which bond atoms to make molecules exists as quantum particles of finite size and intensity.

The bonds that get broken are made of quantum particles of energy that exist as lines of force. Lines of force are made of three types of energy particles. There are photons that are themselves an assembly of a quantity of fundamental attraction energy charges (existing as quantum particles). Each photon assembly is encapsulated with repelling energy charges. A third energy charge produces an affinity between the attraction and repelling charges to produce encasement to contain and prevent radiation of the attraction energy charges. Photons are emitted by protons in a streaming line that are then encapsulated with the repelling charge component of fundamental energy to prevent radiation of the photons.

The repelling charges are attracted to and encase the attraction charges, which means fundamental energy has three components. One energy component attracts like attraction energy components. Another energy component repels like repelling energy components. A third energy component produces an affinity between attraction and repelling charge components. Photons released from the fragmented lines of force that bind atoms radiate from the point of release at the speed of light. Radiating photons are not visible and do not illuminate until their radiation is blocked upon impact with pigmented atoms or molecules.

Upon impact, the attraction energy charge is separated from the encasing repelling energy charge, which in turn releases the energy that produced the affinity bond between the attraction and repulsion charge components. Repelling energy charges (electron charges) are separated from the attracting energy charges and are taken up by the material upon which they are impacted to produce ionization. The attraction energy charges are taken up by the protons and proton power centers of the atoms of the material upon which they are deposited. As a consequence of taking up the released attraction energy charges, the power intensity of the involved atoms that is available to produce new protons and lines of force is reduced by a comparable amount. That diminishes the attracting power intensity of the photons and lines of force produced by the involved atoms, and that reduction in attraction power produces the effect called heat. It is the affinity charge released when the repelling and attracting energy components are separated that produces illumination. The affinity charge is visible and makes the material upon which it is deposited visible. The eyes perceive the affinity energy charges whose intensity varies in proportion to the intensity of the attraction charges from which they are released. The magnitude of illumination is a function of the magnitude of impacting photons.

To make it clear, the energy carried by photons has three components. There is the attracting component that attracts like attracting charges. There is the repelling component that repels like repelling charges. There is an affinity between attraction and repelling charges whose intensity is determined by the intensity of the attracting charge component. The affinity charge produces the attraction between attraction and repulsion charges that results in a repelling charge encapsulating (surrounding) attracting charges. Photons are encapsulated by a repelling charge. The lines of photons emitted by protons are encapsulated by repelling charges to form lines of force. Illumination is the *effect* powered by the affinity component of fundamental presence energy. In addition to being visible, affinity charge energy can do other work as well, such as driving repelling (electron) charges to produce electric power.

Normally the presence energy exposed after photons impact an object is quickly taken up by the material upon which they are impacted, thereby limiting the concentration of released energy at any single point. However, under certain situations, the concentration of energy can be sufficient as to be itself visible. This is what happens at light sources such as lamps or candlewicks (any flame), even the sun, in which case high concentrations of presence energy is seen as a glow. Radiating photons are not visible and do not produce illumination (light) until radiation is stopped by impact upon pigmented physical matter. With flames such as the candle, lantern, or even the sun, the magnitude of released energy particles exceeds the take-up capacity of the photon power centers of the atoms upon which the photons impacted. As a consequence, there is a concentration of attraction and affinity energy that surrounds the wick, which is then visible, seen as a glow. The energy producing the glow is not radiating and dissipates as it is absorbed (taken up) by the surrounding air. Many of the photons released by the combusting fuel do not impact molecules of the surrounding air and radiate unseen at the speed of light until they impact with matter to produce light. Note that the glow does not radiate but maintains its relationship with the wick as long as the concentration of the attraction energy exists in sufficient concentration to be seen.

Support for the stated process is provided by examination of the accompanying photographs where the candle and the lantern are bathed in sunlight. These photographs show what can easily be demonstrated and observed. The conclusions drawn recognizes that the energy involved in binding atom and atom components is inherent with each elementary particle and exists in three parts—attracting, repelling, and interacting affinity.

1. The flames of the pictured oil lamp and candle make no shadow yet are visible. Sunlight (photons) passes through the flame without impacting sufficient matter to produce a shadow.
2. The lamp and candle are made visible by sunlight and cast shadows. The sunlight does not cause the flame to make a shadow, so sunlight is not what makes flames visible.
3. The wall beyond cannot be seen through the flames, meaning the flames with no pigmented matter block vision yet cannot be seen. That behavior confirms that vision perceives the presence energy that is the flame, but vision does not pass through the flame. It also proves that photons, whether radiated or reflected, are not the instrument involved in vision. If photons were a factor in vision, then vision would pass through flames as sunlight does and objects beyond the flame would be seen.
4. The presence energy of which photons are made has no physical presence (no atoms or pigment molecules) yet upon release from encapsulation is visible, confirming that there is a component in presence energy with no physical presence that our eyes can see.
5. Photons are neither light nor heat but, when impacted on pigmented matter, produce the effect called light (illumination) and, when taken up by matter, produce the effect called heat.
6. It is of particular interest to note that the repelling component is deposited as electrons and that the attraction component is taken up to produce heat, leaving the affinity component to produce vision. It is equally interesting to note that the work done, the effects produced, persist only so long as the supply of photons (ergo energy) is sustained. For those interested, consider the story of Moses, who witnessed a burning bush that did not burn, and the story of Shadrach, Meshach, and Abednego tossed into a flaming furnace but were not burned. It can be rationalized that the presence energy in those instances was not produced by combustion and not taken up by the proton centers of material. It would have been seen but would not produce the effect called heat or destruction by combustion.

This illustration shows that presence energy does, in fact, exist; it proves that presence energy has a visible presence and is capable of doing work; it proves the conservation of energy concept that forms the basis for the anticreation conspiracy to be defective, and it proves that the current theory of light and vision espoused by modern physics is defective as well. It is recognition that flames such as those produced by candles are evidence of creator presence that has the greatest potential for good as explained in the chapter entitled "The Mystery of the Paschal Candle."

CHAPTER EIGHT

The Unwitting Conspiracy

The concepts, understanding, and beliefs relative to physics were developed through the study of creation and have evolved over time as understanding developed, except for one—the law of conservation of energy, developed and adopted at the dawning of modern physics. The law of conservation of energy has not only been embraced over the years since being established, but each new generation of physicists seems to have found a way to more strongly reassert unconditional support.

Essentially as developed, the law says, in an isolated system, the amount of energy remains constant. In an isolated system, energy can be changed in form or location but can neither be created nor destroyed. The key operative term in the law is in an *isolated system* that was not and remains undefined by the law. Also, an inventory of energy was not taken at the time, so there was and continues to be no proof that the law is legitimate, and it failed to establish where the energy that existed at the time of creation came from in the first place. The law does, however, acknowledge that the *power* to do work is inherent in matter and available for use directly or through conversion to another form, but it failed to establish where the energy comes from that conveys that power.

It is important to get the horse in front of the cart. The *universe* is the isolated system and everything else in the universe, including the energy to convey power to matter to do work, is a component of that system. The unproven concept of conservation of energy was interpreted to consider each subassembly of the universe as an isolated assembly, which they are not. That interpretation became so important to the anticreator beliefs of the community of scientists that a new type of mathematics was developed to prove that interpretation. The conclusions of Galileo, Sir Isaac Newton, Albert Einstein, and others of their time and since would have been significantly affected had the law of conservation of energy been properly interpreted.

Belief in the conservation of energy theory has been so strongly and blindly accepted that it is taught throughout academia worldwide with the deliberate exclusion of any other consideration. That unproven belief has become a conspiracy to control the mind of man, whether willfully or unwittingly, with anticreation, antireligious connotation. It has caused and is causing serious harm.

As a means of complying with the law of conservation of energy, a conclusion developed that the motion of objects or effects produced by the expenditure of energy is energy. The conclusion that motion, force, and effect can produce work is easily demonstrated and is not argued. However, when such conclusion excludes the obvious fact that energy had to have been expended to produce the motion or effect, the conclusion is defective just as is the original conclusion that accepted the power to do work as inherent in matter without establishing how that power became inherent. That precisely is the point being argued. There is no problem with arguing that religious beliefs on the subject are faulty, but it is patently wrong to stand on the conclusion that the power to do work is inherent in matter without explaining the form of that power and how it is sustained and certainly without entertaining alternative or clarifying explanation.

The concern over a conspiracy to control beliefs is not so much that such will produce adverse behavior of nature or that it will result in wasted time, talent, and capital chasing wild theories, although with a limit on time, talent, and capital, it would be better that they are not wasted chasing misdirected, biased theories. The concern is of the impact on the behavior of the human race being subject to control over mind-sets, thoughts, and beliefs—in a word, brainwashing.

There are two glaring examples of the distortion caused by the current belief. Foremost is the belief that gravity is a form of fundamental energy wherein *fundamental* is inferred as existing without cause. As illustrated in chapter 1, it does not take much to show that gravity is caused by the *energy* inherent in the fundamental, the elementary particles involved in the creation of matter. A second example that demonstrates the seriousness of this unwitting conspiracy was experienced recently while trying to interact with a claimed physicist following a statement that the sun generated the energy it displays.

There is no argument that a lot of energy is displayed by a sun, but such a statement is glaringly misleading when, in fact, every single quantum of *energy* existing on and in the sun is supplied to and through the matter of which the sun is made. The sun is not an isolated system. It is a part of the universe system. The kinetic form of energy produced by the components that make up the sun cannot sustain that behavior if for no other reason than a considerable quantity of those components are not only expended to do work but are radiated out of the sun system. Yes, it is rationalized that, eventually, after several more billion years, the sun will use up its energy-producing material and the sun, a star, will die. There is evidence that stars die, but the evidence indicates that dead stars are extremely dense assemblies of massive amounts of matter, which infers a cause other than running out of matter.

As presented, it appears at first glance that with acknowledgment that the power to do work is inherent in matter, there is no argument. That is not the point. The argument is over the nature and source of that power to do work and how that power is sustained since it defies all logic to blindly (on faith) accept that energy is never expended or lost from a subuniverse system or that an initial supply of energy has remained unexhausted for the millions of years since that material was created. An outline of the mechanics of a process that could develop, deliver, and sustain the energy to power creation is provided in chapter 2 and, as explained, does not violate the unproven law of energy conservation. It may be too much to expect that some effort could be put into trying to prove or disprove the concept offered and not just dismiss it because it closely parallels the religious belief excommunicated by the law promulgated in the nineteenth century that is being interpreted to provide cover for a conspiracy to denigrate the concept of a creator.

The conspiracy founded upon the law of conservation of energy is only one side of the abyss that exists between the beliefs espoused by science and Judeo-Christian-Islam beliefs. Thorns were injected into the world of men by the disinheritance of Ishmael along with the beliefs that the Hebrews alone were God's chosen and that the promise to make Abraham the father of many nations meant an unending sovereignty over a territory. It can just as easily be argued that the Hebrew people were chosen as the instrument by which God's will was to be manifest to all mankind and that the establishing of civil rule was not intended. At least those arguments seem to be supported by the Word of Jesus. At any rate, those beliefs do not interfere with the ability to understand fundamental creation physics. Carte blanche belief of the version of creation set out in Genesis 1:1-26 without consideration for the fundamental physics involved adds credence to the argument of the anticreation advocates, but recognition of the fundamentals of creation physics adds credence and clarity to the genesis explanation.

The adverse effects of the anticreation conspiracy on physics principles can be documented, and in that regard, erroneous conclusions that result will have no effect on creation, only on the understanding of creation. There is concern for the waste of time, talent, and capital that accrues from pursuing erroneous beliefs, but the greatest concern is for the adverse effects on the youth of the world as they struggle through adolescence to understand and develop beliefs needed to guide their lives. The struggle is made extremely difficult with the mandate that only the anticreation beliefs developed by conspiracy-driven factions can be taught in schools around the world.

Clarification of the fundamentals of creation physics provides a solution for the feud of beliefs being waged at the expense of the adolescents engaged in developing beliefs of their own.

CHAPTER NINE

The Mystery in the Paschal Candle

Candles have had a traditional place in religious activities since the beginning of religious worship. Tradition generally holds that the candle is a symbol of the presence of the creator God. By analyzing the physics principles involved, it can be shown that the flame of a candle is more than just a symbol. Flames are, in fact, the energy of creation, the presence of the creator God as explained by the experience of Moses when confronted by a burning bush. Physics explains how the command "Let there be light" translates into the energy of creation, the energy revealed by a candle flame.

Evidence supports the conclusion that what exists now in the universe is the result of creation that in religious belief began with the command "Let there be light," and as claimed by physicists, it was the big bang. In either case, creation began with the application of energy, leading to the conclusion that matter is made of energy. That conclusion is supported by the equation $E = mc^2$ where E = energy and m = mass and c is a multiplier to change from energy to mass units of measure. Since only the creator existed at that time, it leads to the conclusion that the energy responsible for what has been created is the presence of the creator. Chapter 2 provides an explanation for how a presence produces energy. The Genesis explanation of creation is understandably an outline that does not provide a detailed explanation of the fundamentals involved, and physicists have either deliberately or unwittingly chosen to disparage rather than explain the physics fundamentals of the creation process. As a consequence, the fundamental physics principles pertaining to creation have remained a mystery.

Accepting that creation began through application of presence, the first order is to understand how energy was transformed into matter. In other words, the creation of a fundamental building block, a particle with physical presence. The argument is made in chapter 1 that energy with both a self-attracting and a self-repelling component, as advanced by ancient Chinese beliefs as yin-yang forces, will produce a physical presence. Efforts to provably identify such a fundamental particle have been unsuccessful. However, neutrinos possess characteristics understood as needed by a fundamental particle.

For neutrinos (fundamental particles—God particles) to function, to produce attraction and repulsion forces over time, requires that the power to sustain those forces would have to be inherent in each fundamental particle individually with provisions for supplying energy expended in maintaining those forces. To serve as the fundamental particle of all creation, the behavior of fundamental particles would have to be absolutely exact in all respects with interaction limited to like particles and like forces only. Each such particle would contribute one unit of self-attracting power, one unit of self-repelling power and one unit of attraction-repelling affinity power. Since there is no reference or physical presence against which to measure, there is no known way to prove the existence of presence energy except as will be illustrated visually and through recognition of the forces produced. The *force* produced by the *attraction component* of the presence energy that powers neutrinos is known as gravity or gravitational attraction. Gravity exists as an absolute throughout creation. It is detectable, capable of being measured, and is self-proving.

The creation of protons provided the means of transforming the basic fundamental presence energy to enhance power and flexibility. The behavior of protons is absolute, but their behavior provides the ability for flexibility. Protons are the second-generation building block and energy system by which atoms are created and, as a consequence, all higher forms of matter and energy including life. Protons transform the presence energy sustained in, carried by, and supplied through elementary particles into lines of force with which atoms and all subsequent matter is assembled and powered. Protons take the presence energy attracting, repelling and affinity charges from the elementary particles of which they are made and reassemble them into continuously emitted lines of force whose power to attract and bind is variable to support flexibility and a variety of functions.

The *attraction* charge component of presence energy, called gravity, supplied through the elementary particles of which protons are made, are repackaged into packets of 1,840 units called photons. The *repelling* charge component of presence energy, called electrons, remains individual. The continuous stream of photons emitted by protons is surrounded in the process of emission by the repelling electron charges to form lines of force. The stream of photons, with their self-affinity attraction, gives lines of force the power to constrict, and that, in turn, gives the power to bind atoms and assemblies of atoms. The primary purpose of the surrounding repelling electron charges is to encapsulate and prevent the streaming photons from radiating. The encapsulating electron charges also produce electricity and magnetism, a separate subject. Without the constraint of the surrounding electrons, the photons would radiate at the speed of light. Binding atoms and assemblies of atoms is but only one of a number of things lines of force do, and lines of force are frangible.

The energy that produces heat is the number one enemy of lines of force, yet the photons that are components of the lines of force emitted by protons are the carriers of the energy that produces heat. When the protons that produce lines of force take up residue following the expenditure of photon energy to produce illumination, the power of the photons produced by a proton is diminished. Weakened photon power weakens the binding power of the lines of force of which they are components. Weakened binding power causes atoms and molecules to expand and, when sufficiently weakened, causes the lines of force to rupture and the photons involved in making that line of force radiate. That is the process involved in combustion, explosion, solar reactions, and the production of light. Released photons radiate at the speed of light. They are not visible and do not produce illumination until their radiation is stopped upon impact with pigmented matter. It is the attraction component of presence energy that are released upon the matter on which the photons impact. With sufficient concentration, presence energy is visible and, by that means, makes the impacted material visible as long as the energy of photons continue to be released. That process is called illumination.

The attraction component of presence energy deposited upon the impacted material is quickly taken up, absorbed, by the material, and produces the effect called heat, as previously explained. The retained heat energy causes more lines of force to fragment, releasing more photons that release more heat energy upon impact with other matter; this process is called combustion. That is the process involved with a candle. However, the material of which candles are made is highly combustible, capable of releasing significant numbers of photons. The air surrounding the candle (hydrogen, helium, oxygen, and water vapor), while of sufficient presence to block some of the radiating photons, does not have mass sufficient to rapidly take up the released presence energy. As a consequence, there is an accumulation of presence energy around the point of combustion (the wick) that is seen as a glowing flame. The glow does not radiate but dissipates as the concentration of presence energy is reduced through transfer to the cooler surrounding air.

The accompanying picture of a candle illuminated by sunlight illustrates the physics involved. Candles block the photons of the radiating sunlight making the candle visible and causing a shadow to be cast on the wall beyond. The flame does not cast a shadow, so it is not blocking the sunlight's photons, so it is not illuminated by energy released by impacting photons. The flame is not an assembly of atoms; it  is not matter but is visible. The picture of the candle illuminated by sunlight proves vision involves perception of presence energy that not only illuminates the candle but is the glow seen as the flame. Furthermore, since the wall beyond is not visible through the flame, photons that pass through flames are not instruments of vision, adding proof that vision is perception of presence energy.

Presence energy came into being with the command "Let there be light." The glow of a lamp is, in fact, the presence of the creator—God to those so inclined to believe, but a provable almighty presence in any case. The energy seen as the glow of a candle is the same energy that engulfed the bush that did not burn as seen by Moses and that produced the flaming furnace that engulfed but did not burn Meshach, Shadrach, and Abednego. In those instances with the Presence present, combustion was not required to release presence energy.

CHAPTER TEN

Fueling the Sun

There have been a lot of theories advanced attempting to explain how the sun works, what fuels the sun, and how long the sun's fuel supply, and therefore the sun, will last. The generally accepted theory is that the sun is primarily hydrogen atoms with the fusion of hydrogen into helium being the reaction that causes the sun to do what the sun does, the production of photons and thereby heat and sunlight. Current theory is that the supply of hydrogen finite and at some point in time will be exhausted and the sun will stop working. Recent speculation estimated 620 million tons of hydrogen is transformed into helium each second, and the supply of hydrogen will be exhausted in about 5 million years.

Earth to Scale

Photo Courtesy NASA

That theory reflects the equivalency principles set by the law of conservation of energy that considers the sun to be a system wherein energy and mass are neither created nor destroyed but, rather, are convertible. Current theory embodies the belief formed after the success of the hydrogen bomb, that the energy released on the sun is the product of fusion and fission nuclear reactions, where helium is converted to hydrogen (fission) and hydrogen is reconverted into helium (fusion) with energy released in the form of light and heat by both reactions. The conclusion that the sun's fuel supply will be exhausted does not follow the equivalency theory and has not been explained. That, along with other inconsistencies in theory, indicates that the whole explanation for the sun's behavior is defective.

The sun is the center of the solar system, an assembly that consists of considerable matter (mass) created from atoms that are, in turn, created from elementary particles. Therefore, the center of gravity of the entire solar system, an assembly of matter having mass, is at the center of the sun and conveys a considerable concentration of attraction energy that produces considerable gravitational attraction, a force. The entire attraction energy intensity of the solar system's center of gravity acting from the center of the sun establishes and maintains lines of force with each elementary particle of matter in the solar system beginning with the elementary particles of the sun itself.

The force applied by the sun's center of gravity to attract and bond elementary particles is diminished inversely as the square of the distance from the center of gravity to each specific elementary particle. Therefore, the intensity of the force of gravity on the elementary particles of which the sun itself is made is sufficient to cause failure of inter-atom bonds. As a consequence, inter-atom and intra-atom bonds are fragmented with the larger atoms the most vulnerable. As bonds fail, the energy that was sustaining those bonds is released as photons. When photons impact matter, the energy they carry is released to produce the effect called light and heat, and the heat released inside the sun and taken up by intact atoms precipitates a chain reaction of bond failures (fission) that continues until the only matter that remains intact, especially at the core of the sun, is matter that is more resistant to bond failure. Elementary particles, protons, hydrogen atoms (principally protons), helium atoms, and carbon atoms are the matter most resistant to fission. Those are the elements that make up the sun's core and principally throughout. As the lighter elements succumb to fission, the amount of fissionable material decreases, meaning there are fewer bonds are being fragmented so that less heat-producing energy is released (fission reaction subsides), and the residual heat diminishes to the point that the core of the sun is sufficiently cool to allow the reestablishment of intra-atom bonds or fusion. Fission stops when the remaining material has no bonds to fragment, leaving protons and elementary particles, both of which, without repelling charges, compact to a highly concentrated mass. In fact, it is conceivable that the core may include carbon atoms in diamond form.

Recognize that the demarcation marking the reduction in fission is not abrupt. There is a transition zone outside the sun's core where a combination of interactions can take place. It is in that zone that bonds can be established to reform helium atoms (fusion). Contrary to belief by at least a significant number of physicists, fusion does not release energy. That argument was made in an earlier chapter. Bonds that form and the maintenance of force by bonds consume energy. The energy that is expended to form and maintain the force of bonds is supplied by and through the elementary particles. However, through fusion helium atoms are formed. That reaction replenishes the supply of helium atoms that are then available to fission and release light and heat-producing energy, thereby continuing to "fuel" the sun's energy-releasing action. Therefore, the indication is that the sun will continue to function as long as elemental processes supporting the fission-fusion cycle remain viable.

One of the things protons do is attract the attraction component energy released by photons. In the process, the power intensity of the LMF radiated is reduced, that is, consumed in attracting the released heat-producing energy. The one thing that could bring an end to the sun's viability would occur if a preponderance of the sun's protons (hydrogen atoms) retained such an amount of heat energy that the intensity of their center of proton power was weakened to such a degree they no longer had the power to produce electron bonding (lines of magnetic force). If that happened, the fusion of hydrogen atoms to form helium atoms could not take place, and there would be insufficient electron bonds (LMF) to break to produce photons, and the fusion cycle could no longer function.

The process by which that cataclysmic event is prevented is complex and not fully understood; however, it appears to involve two processes, both of which are controlled by the sun's magnetic force fields. As explained in an earlier chapter, magnetic force fields contain the energy that produces the effect called heat, but magnetic force fields are subject to leaking and fracture, especially when subject to high pressure. Pressure builds as a result of the expansion that occurs with heating. That is good since pressure is indicated to aid in fusion actions. Some heat energy manages to escape through the magnetic field as solar wind and radiating photons, and the heat energy level is controlled. However, when the heat energy concentration builds excessively, the magnetic force field is blown out and ruptures to release the excess pressure. The blowouts, as shown by the accompanying photograph, are seen as producing the phenomena called solar flares. As a consequence, the magnitude of entrapped heat energy is controlled, and the transformation of presence energy to produce light and heat continues.

Before concluding this explanation, there is an important point to be made regarding the light produced by the sun. Photons released from broken inter-atom bonds radiate from the point of fracture. Many of the released photons escape the sun's interior without striking pigmented material and radiate beyond the sun to produce illumination of other matter and objects such as the earth. However, a considerable quantity of released photons do strike matter (atoms) that are part of the sun, but the material even in the outer structure is heat laden so unable to take up the energy released from impacting photons. The energy released by impacting photons does not radiate, resulting in a concentration of the released energy. The energy released when photons impact is visible, as illustrated in the chapter that explains vision and light. As a consequence, the concentrated energy glows. The glow does not radiate but dissipates through convection action because it is heat. The glow is not pigmented matter as illustrated in the chapter explaining light and vision, so it does not block radiating photons. It is the glow that is seen as the sun and stars in a night sky. The glow is not radiated light, which is not visible.

This explanation leads to an as-yet-unproven conclusion that the solar wind is primarily elementary particles that have been stripped of the affinity charge and the charge that conveys the power to attract. In other words, they carry only the repelling (electron) energy charges. As a consequence, that residue with repelling power is dispersed and carried away by the sun's magnetosphere as solar wind.

CHAPTER ELEVEN

Atomic Theory of Matter

Background. The conclusion postulated is derived through reconsideration of data collected and recorded in textbooks and documented scientific observations that culminated in the atomic theory of matter. Reconsideration is based upon an a priori conclusion that, contrary to the kinetic theory of matter and the law of conservation of energy, the elementary particles, and from there all matter, are powered by energy deposited on and made a constituent part of the elementary particles of which all matter is made. The possibility grew from a realization that current instrumentation that reacts to electrical energy is not capable of sensing the energy that powers gravitational attraction, and as revealed by Galileo and Newton research, the energy empowering gravitational attraction would have to be a constituent component of the elementary particles of matter.

Explanation. Matter is made through the assembly of atoms that are elementary particles of matter. Atoms are made through the assembly of protons and neutrons that are bound with the force of gravitational attraction in addition to the lines of force emitted by protons. Neutrons and protons are made of a precise number of like elementary particles assembled by the gravitational attraction force powered by the energy charge implanted and sustained on each elementary particle. (Quarks theorized as elementary in forming neutrons and protons, are formations of elementary particles and binding force that occur in the course of neutron and proton assembly. They are not elementary particles.) As a consequence, a binding force established between elementary particles results in the creation of neutrons and protons, which assemblies acquire gravitational attraction power equal to the number of elementary particles assembled. The power of an assembly to attract elementary particles acts from the center of the assembly where the attraction intensity or gravitational attraction is equal to the number of elementary particles assembled. The number of elementary particles in an assembly constitutes the mass of that assembly so that the intensity of gravitational attraction at the center of gravity provides indication of the mass of the assembly.

Conclusion. During the assembly of atoms, binding force is provided by the attraction component of the energy charge that is distributed to assembled matter through the elementary particles that are constituent components of all matter. A proton, for example, acquires its power to function from its constituent elementary particles. When bonds are fragmented as in combustion or nuclear fission, the involved elementary components are unbound, and energy no longer used to produce binding force is radiated from the point the bond is fragmented as photons. The energy of which photons are made produces illumination and the effect called heat. The repelling charge that encapsulates the lines of force emitted by protons can be converted to electrical energy by generators or chemical action.

Supporting argument. Three areas in this modified theory stand out as deviations from the current atomic theory of matter.

> First is the explained role and processes of protons. The explanation given was derived after long and thoughtful consideration of proven behavior, including behavior that would be needed to do what protons do. That is, develop and emit energy in a form and with the capacity to bind atoms and molecules to create all the various forms of matter as well as in the form involved in producing light, heat, magnetic force, and electricity. (A more definitive explanation of proton behavior is provided by chapter 3.)

> Second is the recognition that, rather than molecular motion and conservation of energy, an energy source exists (as explained in chapter 2) that supplies the power involved in creation and the sustenance of creation. It is generally agreed, without any fanfare, that celestial bodies take a spherical shape in response to the force of gravity acting from a center of gravity. That is especially evident for earth, where the oceans and waterways are kept in place by the force of gravity acting from the earth's center of gravity. The

tides produced on the earth's seas by the gravity of the moon shows that other bodies develop a center of gravity with the power to attract over distance. That gravity is responsible for spherical shape of celestial bodies is generally just taken for granted, but that phenomenon is speaking clearly, providing indisputable proof that the force called gravity is in response to energy acting from and through a center of gravity that is established in each body (assembly) in the course of assembly. The role of gravity is not so clearly evident in smaller assemblies like raindrops or elementary particles, but the message sent by larger bodies like the earth, moon, and solar system speaks for the little guys. The formation of a center of gravity means that it is the power of elementary particles to attract elementary particles that establish the center of gravity in all assemblies. Raindrops form into spheres with surface tension because of the force of gravity acting from the center of gravity of the raindrop.

Third is recognition that the law of conservation of energy is defective. The atomic theory of matter has been rather conclusively proven; however, some of the explanations of the processes involved are illogical or stand on hypotheticals which, as explained, proceed from the need to accommodate unproven archaic beliefs. By recognizing the existence of creation powered by energy supplied through particles elemental to all things, the physics involved is clearly obvious, obviating the need for hypotheses and mysteries. It gives relevant meaning to what scripture attributes to Jesus: "Everything that is now hidden or secret will eventually be brought to light. Anyone who is willing to hear should listen and understand! And be sure to pay attention to what you hear. The more you do this, the more you will understand—and even more besides. To those who are open to my teaching, more understanding will be given. But to those who are not listening, even what they have will be taken away from them (Mark 4:22-25, NLT).

It is not the purpose to preach or justify religious beliefs, but since physics and religious beliefs derive from the one creation event and since religious beliefs have been factors in the formation of anticreation beliefs embedded in physics beliefs, recognition is given to show commonality of purpose. All should recognize that all Abraham, Jesus, and Muhammad advocated was for belief in what is evidenced by creation (as was so clearly stated in Romans 1:20-23, NLT) quote: "From the time the world was created, people have seen the earth and sky and all God made. They can clearly see his invisible qualities—his eternal power and divine nature. So they have no excuse whatsoever for not knowing God." The scripture quote is made to clarify that the intent from a religious perspective was to gain recognition of the glory of creation, which must be admitted as glorious whether or not one believes it to be the work of a creator. Creation is an event. That creation is the work of a creator is a belief. That does not change the processes involved, which physicists study in an effort to understand, and in that regard, all that is needed is that facts stand free of a biased mind-set.

PART II

Examples of Creation Physics Power

INTRODUCTION II

Part 1 attempts to explain the fundamental physics principles revealed when a source of presence energy is recognized as powering creation. Part 2 documents blogs that have been published to provide an explanation in layman's language to counter textbook explanations of physics phenomena that were flawed and with anticreation bias. Over time, a realization developed that the anticreation bias of physicists was fed by beliefs inherent in the laws and principles that physicists had developed, which direct academia and research. A realization also developed that some beliefs and conclusions developed and adopted as laws in the infancy of physics perhaps unwittingly established an anticreation conspiracy mind-set that blinded understanding of fundamental physics principles that are hidden in the mysteries surrounding creation. Beliefs inherent in the processes employed in creation are presented to show how realization grew that current textbook explanations fail to explain fundamental physics principles. It was in the course of blogging in an attempt to show the errors and flaws in current physics beliefs that a realization grew that there is an anticreation mind-set built into the beliefs and principles that guide the community of physicists.

Eventually the awakening brought on by blogging developed a realization that archaic beliefs that hid understanding were never challenged, purged, or updated in line with newer findings. Rather than let the new findings lead where they may, conclusions were developed to accommodate archaic beliefs. Case in point was the kinetic theory of matter and the law of conservation of energy and mass. When the concept that energy is supplied and distributed to power the behavior of creation, the doorway opened to understand and explain fundamental physics. The blogs included in part 2 illustrate the process of awakening.

BLOG ONE

Flawed Fundamental Physics

Gravity. Through demonstrations at the Leaning Tower of Pisa, Galileo showed that all matter is an assembly of like elementary particles and that the earth's force of gravity acts upon the like individual fundamental particles of an assembly and thereby the assembly. Realization of that fact so important to understanding fundamental physics and the processes of creation has remained hidden all these years. The Leaning Tower of Pisa demonstration proved that it is the number of elementary particles of which an object is made that determines the force of gravity, not physical size, type of material, or density of the material of which an object is made. From that, it is concluded that all matter is created by the assembly of absolutely like particles that are elementary to the creation of matter and to establishing the force of gravitational attraction. Furthermore, the Galileo demonstration proved that each elementary particle is endowed with a precise magnitude of power to attract other like elementary particles, and as subsequently demonstrated by Charles Coulomb, the like charges that power attraction also endow the power for like-charged elementary particles to repel each other. From there is drawn the conclusion that the elementary particles of which all matter is made are endowed with the power to attract and the power to repel like-charge particles or as postulated by the ancient Chinese, yin-yang forces. The multitude of like elementary particles with which the universe was created, existing ubiquitously throughout the space that would become the universe, constitutes an immense energy potential so long as the like charge is maintained, in which the potential is sufficient to sustain the charge on the elementary particles as long as the like particles exist. That theory is justified by the conclusions drawn by

Charles Coulomb and the numerous others who have done research on and have contributed to the understanding of electricity and electromagnetism and the behavior of matter subjected to a charge potential.

Conclusion. All that exists in the universe is produced through the assembly of absolutely identical particles with like charge and behavior that are fundamental to the creation of all things. It is sustenance of the charge on the elementary particles that conveys to elementary particles the energy that powers the behavior of matter. It is the power of mutual attraction between elementary particles and, in turn, between assemblies of elementary particles that establishes the force of gravitational attraction and the characteristic called resting mass, which is an absolute. Understanding fundamental physics requires that the absolute not be contaminated by including other relativistic forces within that classification. It is the charge sustained on elementary particles that is transformed by protons to produce photons and the energy system that powers the electromagnetic form of energy from which light, heat, electricity, magnetism, and weak and strong binding forces are produced.

Light, vision, and illumination. Merriam-Webster defines *light* as, "(1) The act or power of seeing: sight. (2) The special sense by which the qualities of an object (as color, luminosity, shape, and size) constituting its appearance are perceived through the process in which *light rays* [italics added] entering the eye are transformed by the retina into electrical signals that are transmitted to the brain via the optic nerve."

The *Merriam-Webster* definition is quite definitive but for one aspect. It fails to define *light rays*. The current understanding is that light rays are radiating photons as the light rays known as sunlight. Sunlight involves the radiation of photons, and photons do produce illumination following impact on pigmented matter wherein matter is inferred to be made of atoms that are an assembly of electrons, protons, and neutrons that are bound into assembly by lines of force. The argument that follows will show that the light rays that enter the eye are not in the form of photons.

Sunlight that involves the radiation of photons is a form of electromagnetic radiation. Photons are made of the energy that makes vision possible, but photons are not visible, and they do not produce illumination until their radiation is blocked upon impact with atoms that produce pigmentation. Following impact, the energy radiated as photons is visible; it illuminates the matter upon which it impacts and makes that matter visible as long as radiating photons continue to impact. However, the illustration that follows will provide evidence that the light rays that enter the eye are not in the form of radiating photons.

The accompanying photograph shows a candle illuminated by sunlight (radiating photons). The sunlight makes the candle visible, and because the candle is made of pigmented atoms, it blocks the sunlight, causing a shadow to be cast on the backdrop but with one exception. The flame does not cast a shadow yet. It is visible, meaning the photons are not blocked by the flame; therefore, the flame is not made of pigmented atoms, and although the flame is seen to have qualities, its qualities are not made visible by the sunlight.

The photo also demonstrates that the backdrop cannot be seen through the flame; therefore, radiating photons are not the mechanism involved in vision. If that was so, photons radiating from the backdrop would pass through the flame as did the sunlight photons, and the backdrop would be seen.

It is submitted that the flame is the energy carried by the photons released by the bonds fragmented by combustion of the candle wax, which involves the hydrogen carbon bonds and the O2 bonds as they break down to two oxygen atoms. The photons released by combustion of the candle impact the surrounding air molecules, causing fragmentation and release of photon energy. With the air molecules being of little mass, the energy released exceeds the ability of the involved air molecules to take up the released photon energy. As a consequence, there is a concentration of energy that is visible as a glow, the color of which reflects the wavelength of the involved energy. The glow dissipates as the energy, which is now the energy that produces heat, is taken up by the surrounding air molecules.

Conclusion. Radiating photons (sunlight) are not light but the energy that produces the effect called light, which is illumination. Radiating photons are not the mechanisms involved in vision. It is the energy that powers elementary particles, otherwise defined as presence energy, supplied from the singularity that encompasses the universe that is released as illumination that our eyes are equipped to perceive.

Summary. Current theories regarding the fundamental physics principles especially as associated with the conservation of energy, gravity, electricity, mass, light, and heat are flawed.

BLOG TWO

Looking at Vision

Do not let the simplicity of the physics of vision close your eyes. The processes involved with vision are in the realm of miraculous, but the physics involved is made simple when analyzed using fundamental physics principles. Quite clearly and simply, vision is accomplished by perception of objects by the eyes, and contrary to modern physics theory, the transmission of *photons* from the objects in view to the eyes is not involved. That conclusion is also supported through consideration of night vision as shown by the photograph taken using night-vision technology. The detail in this photo is quite clear. An explanation of how night vision works claims that energy *radiated* from the object (which means photons) is captured and amplified by the night-vision camera. Sounds logical except for one thing. Radiated energy sprays in all directions, in which case the size of the object would increase exponentially with distance and all the camera would see would be a blur; there could be no detail. It should be simply obvious that the process of vision does not rely upon energy *radiated* as photons, but that vision is the perception of another form of energy. The

Photo Public Domain,
Night Vision Effect Courtesy photoandvideograpy.com

eyes and a camera are sensitive to the energy that illuminates the view and registers what is perceived. The energy that illuminates is the energy released when radiating photons impact pigmented matter, which is the attraction component of fundamental energy, and the attraction component of fundamental energy is presence energy. Stated simply, it is presence energy that is involved with vision; it is not radiated photons.

There is a second type of night vision (thermal imaging) although that process is incapable of revealing detail. It relies upon sensing the energy that produces the effect called heat. An object is made relatively hot when the protons of which it is made take up the energy that produces the effect called heat. The energy that produces heat is residual illuminating energy (light) and is visible, perceivable, by our eyes. (That is why hot metal glows.) The intensity of the retained energy varies depending upon the strength of the bond from which it radiated. Our eyes are sensitive to variations in intensity of retained energy. As in the case of normal vision, what is seen is perception of presence energy that does not rely upon the radiation of photons from the object viewed.

Added Comments

The conclusion derived through logic in this instance constitutes an entirely new revelation and deserves recognition as such. Vision is line of sight and is blocked by intervening objects, including concentrations of presence energy. Subsequent to the posting of this blog has come the realization that the energy our eyes perceive is the same energy that has been given the name *graviton*. Gravitons are the attraction component of presence energy, and presence energy acts instantaneously, so vision is instantaneous. That means with vision there is no time delay while photons travel as there is with the transmission of light. That, in turn, means that it does not take ninety light years to see a star whose radiated photons take ninety light years to get here if, in fact, any such light ever does get here. That raises a serious consequence. Since it is not the energy radiated as photons that we see but is, rather, the glow of presence energy, which is instantaneous, how can the speed of light be useful in estimating interstellar distance?

BLOG THREE

A Chain Molecular Reaction

This is a theory developed using the principles of fundamental creation physics. It is understood that the energy involved in the formation of storm cells is heat with the heat energy acquired principally from warm tropical waters. It is also rationalized that storm cells develop when warm, humid (water-laden) air collides with cold air and that the intensity of storm cells is maintained because of a continuing supply of warm, humid air.

That is a sound theory; however, that explanation leaves two troubling questions. First, why is the warm, humid tropical air not happy to just unload by condensing to produce a nice tropical shower? Second, is it just the continuing movement of warm, humid air to meet colder air that accounts for the development of prolonged destructive-level storm-cell energy? There is an answer for both questions. The water molecules that make air humid expand when heat energy is taken up (absorbed) and contract when heat energy is given up. Warm water vapor (clouds) is less dense, allowing a greater quantity to reside in the atmosphere in the manner of a ship floating in the sea. When humid air is cooled, the water molecules shrink (condense), increasing density to the point where the force of gravity overpowers the lifting force of the atmosphere and it rains. The process of condensation involves the transfer of heat from the water molecules to the surrounding cooler air. The making and breaking of intermolecular bonds (chemical reaction) is not involved until expansion exceeds the limit of bonds involved in binding atoms, causing bonds to fragment.

When intermolecular bonds fragment, the energy that produced the bonding force is released as photons. That energy is fundamental energy supplied through the fundamental particles of which the involved water molecules are made. It is not heat energy carried by the warm water (vapor) molecules. It is additional heat energy carried by the fundamental particles of which the molecules are made and released when intermolecular bonds are broken, which explains why storm cells intensify beyond the nice tropical-rain shower stage. The radiated photons impact other water molecules where the additional heat causes expansion, bond failure, and the release of additional heat-producing energy. The cycle repeats over and over, ever increasing in magnitude in the manner of a chain reaction. The process involved in transforming gentle tropical rain showers into dangerous thunderhead storm cells constitutes what has been termed a chain reaction akin to an atomic and nuclear reaction but involving molecule binding rather than atom or nucleus binding.

The severity of atomic and nuclear chain reactions is managed though the use of control rods, which introduce matter that does not react (an impurity) and which can extract heat to keep it from running away or melting down. At the present time, prevention of storm-cell meltdown relies upon naturally occurring controls. The reaction material (warm, humid water vapor) is not pure, and the resupply is not always consistent, and storm cells are at the mercy of self-destructive winds. Nevertheless, storm cells entail significant amounts of uncontrolled chain-reaction energy.

The intensity of a storm cell can be controlled in the same manner as the intensity of atomic and nuclear reactors is controlled, except the control mechanism must be mobile. One possibility involves developing a carbon-fiber tube configured to be an electromagnetic coil to extract heat energy out of high-energy clouds with the earth as a heat sink before the chain reaction gets excessive. It may be feasible to erect fixed control tubes in tornado-alley locations like Oklahoma and Kansas, where severe storm cells are prone to develop.

Added Comments

The technology necessary to effect storm mitigation is explained in chapter 5. Magnetic force has been in use for years to contain nuclear fusion and fission where there is considerably greater pressure and temperature than exists in thunderstorms.

Storm mitigation as proposed is indicated to be environment friendly. Dangerous and unruly chemicals are not used. Weather need not be adversely altered. The amount of heat energy extracted can be limiting to that necessary to prevent destructive storms, leaving the weather system intact.

Since storm mitigation has the potential to reduce damage, the funding of research and development should be of interest to the insurance business as well as federal and state governments.

BLOG FOUR

Intelligent Design

Merriam-Webster defines *intelligence* as "(a) the ability to learn and understand: to reason. (b) The ability to apply knowledge or the act of comprehension. It defines wisdom as (a) accumulated learning, knowledge, (b) insight, (c) good sense: judgment." In simpler terms, maybe both involve recognizing that the horse goes before the cart, which, interpreted, means intelligence existed before creation. One thing is for sure—natural selection and evolution could not have directed the manufacture of the fundamental elements of creation, and whether or not imported from another universe or kingdom, intelligence and wisdom were prerequisites to that first step in creation. It may be argued that after the enormous collection of presence that became all things, assembly would follow of its own accord except that the presence particles would have had to know the need for and know how to establish opposing attracting and repelling charges that would serve all subsequent needs. Of course, such an argument ignores accounting for the origin of the presence material. That argument and all subsequent arguments are illogical and incomprehensible when intelligence is not recognized as a prerequisite for every step in the creation process.

Reverting to the *Merriam-Webster* definitions, there can be but one conclusion: creation encompasses all wisdom and knowledge. We, having been endowed with an intelligence capability in the likeness of the creator, are able to gain knowledge and wisdom by studying and learning the processes employed in creation. From there, anything is possible as long as the horse is kept in front of the cart.

Arguing that evolution (progression of creation) is a result of natural selection is shallow and a disservice to the mental capability of human beings. Modification of behavior in response to environmental changes would be expected with an intelligence-directed creation. The natural selection argument fails to establish where the intelligence comes from that directs change. The natural selection mind-set became established because of the anticreation conspiracy set in place by the law of conservation of energy adopted by the community of scientists that ignores a creator-driven creation outright. An intelligent mind cannot accept that a viable modification would result by chance. An intelligent mind would recognize that an intelligence-driven change is far more likely to produce the needed results.

The argument by religious believers that creation was and is a religious act is damaging to the concept of religion. Granted that belief in creation and a creator is the basis upon which religious beliefs are formed, creation was not in response to religious beliefs. Religion grew because the human mind was able to recognize and respect the greatness of creation. Modern religion grew from the belief that the behavior of humans was not all that it could be and that the creator sent word that adjustment in behavior was needed. To establish authentication upon which to build belief, ties were made to the creator as documented in Genesis and the concept that creation was, is, a religious matter became accepted when, in fact, religion is concerned with the behavior of humans, not the process of creation.

The division of beliefs between the community of science scholars and the religious community evaporates when an energy-driven creation is put at the front of all considerations and beliefs. For one thing, recognition of an energy-driven creation will not change one iota of reality, and what is and what will be will not be changed except for relief from dissention and animosity.

Blog Five

Sonoluminescence

Sonoluminescence is an as-yet-unexplained phenomenon whereby light is produced when bubbles of air in water are ruptured by sound waves.

Water is a collection, an assembly of molecules wherein two hydrogen atoms are bonded covalently with an oxygen atom. In the liquid state, those molecules form additional bonds between the hydrogen atoms; this is called hydrogen bonding. It is that unique bonding characteristic that gives water the ability to take up and release large quantities of heat energy without changing from solid to liquid to vapor state as well as the flexibility to exist in liquid state even when subject to extreme flexing forces. Furthermore, the unique bonding characteristics of water molecules cause liquid water to form a special bonding for those molecules on the surface that are unable to bond normally with other water molecules. The special bonding of surface water molecules is called surface tension and involves more and stronger bonding.

If you collapse an underwater bubble with a soundwave, light is produced, and nobody knows why.

Photo Courtesy Yourdailymedia.com

In the course of existing, there is significant making and breaking of intermolecular bonds, both the covalent hydrogen-oxygen and the hydrogen-hydrogen types. When bonds break, energy in the form of photons is released, but when new bonds are made, the molecules involved take up heat energy, which is the energy carried by photons after release. That is, heat energy no longer exists as protons but as a quantity of the attraction component of fundamental energy. Heat energy is no longer in the form of photons and, when taken up by the nucleus of surrounding atoms, it weakens their power to produce binding power until or unless the heat energy is transferred via convection to adjacent atoms, which is readily done in water. The recurring bond-breaking that takes place with water and water vapor molecules is relatively low-level, so it does not normally produce a visible or a noticeable heating reaction.

Under the right circumstances, bubbles form in water when, by its inherent bonding behavior, the water forms surface tension to surround and encapsulate the air, which prevents dispersion of the air. The interjection of sound waves causes the unique surface-tension intermolecular bonds to fragment. That is a significant reaction in a short time with a corresponding significant release of photons. The released photons impact hydrogen and oxygen atoms in the water molecules that illuminate the water molecules with presence energy that is visible, seen as luminescence.

Added Comments

Without laboratory equipment for a precise determination, reasoning supports the conclusion that the blue tint and low intensity of sonoluminescence is the product of broken hydrogen bonds.

Under the circumstances, creation physics principles should be given credit for producing an explanation where modern physics principles failed. This is possible because creation physics recognizes the role of presence energy in the creation of molecular bonding.

Blog Six

Light-Emitting Diodes in Creation Physics Terms

Photons are the fundamental attraction charge produced by protons that are normally encased by the fundamental repelling charge or electrons as currently understood, in lines of force that bind protons and neutrons to form atoms and atoms to form molecules. Normally photons are released from their electron charge when intra-atom/molecule bonds are broken as in chemical actions such as combustion or as a consequence of electricity flowing along a conductor, especially a lamp filament. When radiating photons contact pigmented mater, the energy they carry is released to produce the effects we call light and heat. Conventional lightbulbs require the passage of a significant quantity of electron charges over a special filament conductor to produce artificial lighting. That requires the application of considerable energy to force electron charges to travel over the high-resistance filament conductor.

Following accidental discovery through considerable research and experimenting, a method of obtaining release of photons using light-emitting diodes (LED) has been developed. The process involved releases photons without the need for expending energy to break intra-atom/molecule bonds. Since the process is technically complex, the current scientific explanation is confusing. This is an effort to provide an explanation in common terms using physics principles revealed through the study of creation physics.

The materials used in making LEDs are specially selected and prepared to achieve the needed behavior characteristics. That is a subject where definition is not necessary in order to explain the process. People in general are quite familiar with the way electricity normally works. A difference of potential is generated and carried over insulated, isolated wiring to the point of use. When a switch is closed, the difference of potential forces electrical charges (electrons) to travel over the conductors and power work. In lightbulbs, the movement of electric charges from atom to atom along a lamp filament, causing lines of bonding force between atom/molecule components to be broken, releasing photons that carry the energy that produces light and heat. The selection of material and construction of LEDs can achieve photon release without breaking intra-atom/molecule bonds. This is a major breakthrough in technology with great promise. It would be nice if people were able to understand the technology involved.

LEDs are constructed of two special wafer-thin materials bonded together to form what has been designated a p-n junction. The materials are extremely poor conductors of electricity, so electric current as such does not flow. The LED is connected to a DC power source so that the difference of potential (voltage) is impressed across the LED. Because of the nature of the two select materials, electron charges from the negative pole of the power source are stopped at the p-n junction, but some minute amount of diffusion does occur. As a consequence, the electrical potential of the material (p material) connected to the positive terminal of the power source is raised to the potential of the positive terminal. That is equivalent to giving each atom or molecule the power to attract an extra proton. The p-material atoms/molecules, now with an extra electrical potential, are capable of taking up an additional electron charge but do not have the extra proton needed to acquire one, and the atom/molecules with the extra proton capability extend up to the p-n junction, with possibly some diffusion into the n material on the opposite side of the junction.

As a consequence, there are some p-material atoms/molecules in the diffusion zone that have the capacity to take up an additional electron charge, and there are free electron charges available in the n material to be taken up, *but* the instant the electron charge is taken up, the receiving p-material atom/molecule no longer has an electron charge deficit, so the energy that gave the p-material atom/molecule the power to take an added proton releases the energy involved as a photon. That p-material atom/molecule immediately acquires the power to take up another electron charge, and so the cycle repeats.

Because the surrounding material is wafer-thin with a lattice structure, most of the released photons are able to escape without being taken up by the atoms that form that material. Ergo, radiating photons capable of producing the effect called light without the need to break a line of force bond as in a tungsten filament. Essentially the only power consumption by an LED is the energy carried by the ejected photons.

Added Comments

Even using nontechnical language, the technical aspects of light-emitting diodes may be puzzling without some basic knowledge of electricity. This blog recognizes the technological breakthrough that produced light-emitting diodes. From the explanations available, it just seemed those who developed the technology did not understand precisely the physics principles involved. It seemed logical that understanding those principles could help advance this technological breakthrough.

The point is, with conventional lightbulbs, energy is expended, overcoming the resistance of filament material to the flow of electricity. With diodes, electricity doesn't flow except to the extent needed to establish and maintain a difference of potential (voltage or charge). When those conditions are properly established, then the force charge established on the diode material allows the molecules of which one of the two materials are made to take up an extra proton while leaving molecules of the other diode material one proton short. Then interaction between the two materials pulls the extra proton, breaking the bond formed when the diode material took up an extra proton. When the extra proton is released, the photons involved in making the bond are released to radiate and do what photons do—produce illumination.

The key to understanding is to realize that it is the presence energy supplied through the fundamental particles of matter that makes the bond that produces the light. It does not come from the electrical power system. The electric power system merely establishes the charge allowing the diode material to take up the extra proton. The proton is not consumed and is not lost. Immediately upon release, the charge is reestablished on the diode material and it takes up another extra proton.

An understanding of this process is possible only if the role of presence energy is recognized, the fundamental principle of creation physics.

BLOG SEVEN

Space, Time, and Gravity

The problem—that is, the place where the train got on the wrong tracks—began when some genius concluded that because gravity and acceleration produced similar reactions, gravity and acceleration are the same thing. Please understand, in acceleration, things in motion are reacting to the application of a force in addition to the force of gravity, so those reactions are not dealing with the same thing. The force of gravity is fundamental to creation. Gravity exists with or without the application of additional forces such as those that cause acceleration. Acceleration and the force potential of acceleration are separate, distinct phenomena.

Consider the example used to explain the commonality of gravity and acceleration: Standing on earth, a boy drops a ball, and it falls to earth (pulled by gravity). A boy on an accelerating elevator drops the ball, and to an observer on the elevator, it falls similarly but in a direction opposite to the force causing acceleration. That reaction is like being pressed back into the seat of an accelerating car. That falling sensation was in response to the force causing acceleration as it appeared to an observer on board the accelerating object. It did not have the same appearance when observed from earth, and they are apples and oranges. To an observer on earth, the ball dropped in an accelerating elevator appeared not to fall but reacted to the force produced by the accelerating elevator, in which acceleration required the expenditure of energy from a source other than the energy that produces gravity. Let's get the train back on track.

This is not to say that forces in reaction to acceleration and motion are not real. The point being made is that force due to gravity and force due to acceleration are not the same. The force of gravity is a fundamental force in response to a specific application of energy and must remain a separate consideration especially when dealing with the physics of creation. It was creation that introduced the processes that produced gravity and, as a consequence, motion, distance, and time. Those things did not exist until after the big bang. The force of gravity is fundamental to creation. It existed before the big bang. Therefore, to understand space-time requires understanding space-time. Offhand that may seem complicated but is, in fact, quite simple when the train is on track.

Gravity is nothing more, nothing less than the force inherent in the fundamental energy of creation. Gravity is the reaction to the affinity attraction charge carried by each fundamental particle (neutrino). Each fundamental particle carries one absolute unit of that gravity-producing affinity charge, and when energized (cosmic background radiation, a separate subject), each fundamental particle acquires a corresponding repelling force. The interaction of the charges carried by each fundamental particle is the force we call *gravity*. Un-energized, the fundamental particles existed at a single point, a singularity. Upon being energized (think of the command "Let there be light"), a repelling force is established, acting between and causing the separation of the fundamental particles in a manner described as the big bang. The fundamental particles, even though expanded, exist as an assembly bonded together by interacting gravity force. The expanded cloud of fundamental particles remains one body that acquires an attraction force equal in intensity to the sum of all the fundamental particles. That is, the universe has a singularity, a center of gravity, from which the energy that produces gravity is radiated and which force binds the universe as one assembly even though existing as galaxies, solar systems, individual planets, asteroids, or free unassembled fundamental particles.

That is the universe following the big bang. Through the fundamental particles, the universe has acquired the power of gravity, but at the same time, the big-bang expansion interjected distance and space (between particles) so that interaction was no longer instantaneous. Interaction required the expenditure of time and introduced a new phenomenon. The introduction of distance means interaction is not instantaneous but occurs over time, which introduces the matter of speed of interaction since instant interaction is no longer possible. The speed at which gravity force acts over distance is determined by the rate at which the energy that powers gravity expands from the singularity where the power intensity is maintained constant and interaction between particles is instantaneous because there is no distance, no space.

Without elaborating, the energy radiating from a constantly maintained singularity expands at the same rate at which the volume of a sphere increases as the radial distance increases. The rate of expansion is determined using the formula volume = 4/3 pi r³ where r = radius, which upon solving translates to a reduction in intensity (power) of the energy being radiated at a rate in direct proportion to the radial distance over which it is acting, squared. The rate of expansion of the presence energy that powers gravity is a constant. It has been measured and determined to be 186,000 miles per second, the speed of light, which is the speed of fundamental energy radiation.

The point being made is that gravity and the behavior of the energy that produces gravity are absolute fundamental creation physics. Gravity does not warp time or space (distance), and time and space do not produce the force of gravity. Those concepts are the creation of the Einstein theory of general relativity formulated in an effort to explain gravity without acknowledging the role of presence energy as being fundamental to creation and to physics, which is a consequence of the conspiracy set in place by the law of conservation of energy.

When the force and motion produced by fundamental gravity is considered as a fundamental absolute, independent of forces applied from other causes such as acceleration or multiple sources of gravitational attraction, then the concepts and beliefs posited by general relativity are exposed for what they are—manufactured explanations to justify disregard for the fundamental physics inherent in creation. In other words, gravity does not distort or warp space and time; rather, reactions to other forces applied to a body will alter gravity's reaction to cause variation in the direction or acceleration or speed that the body moves or would move if free of those influences. Calculation to predict or explain the effect of multiple forces applied to a body would require more than basic math but not the theory of general relativity.

Added Comments

The concept of warped space and time was developed by Albert Einstein in his effort to explain gravity. Gravity is one of the four forces admitted by Einstein as being a mystery. As pointed out in chapter 1, gravity is a *force* produced by presence energy. Gravity is not a fundamental form of energy although the force of gravity has potential (kinetic energy) capable of doing work. The force of gravity would not exist except for the power provided by presence energy. The concept of space-time warping is not proven except mathematically to the satisfaction of a few. The physics involved with gravity is covered in chapter 1.

BLOG EIGHT

The Speed of Time

All that now exists as an environment consisting of the space we call the universe initially existed as a singularity where there was no distance or space and where interaction was instantaneous. Space (our universe) was created with the big-bang expansion. The yin-yang (attraction-repulsion) forces acting between fundamental particles precipitated the big bang. The big-bang expansion created distance so that interaction involving what was created requires the traversing of distance and interaction is no longer instantaneous.

We have given the consequence of traversing distance the name *time* and devised a system for standardizing the consequence of traversing distance into intervals based upon the earth's rotation (a day) and solar orbit (a year), which are consequences of traversing distance, and we called those intervals *time*, thereby establishing a means of correlating the passage of time with the earth's movement in space, a consequence.

Time, by our definition, does not designate speed but, rather, is a factor in designating a rate of speed for distance traversed. Sometimes it is used in conjunction with the distance the earth has traveled. For example, waiting for an appointment or waking up in the middle of the night and waiting for the time to get up are "time goes by so slowly" experiences while not having enough time to get everything done seems time is passing too rapidly. But such experiences are not because the speed of time changed. Time is only a consequence of event passage, and whether it is deemed slow or fast is a perception. Time does not stand alone; it is a factor in establishing a rate of speed, in our case made with reference to the distance the earth travels in space.

Time established by the system of measurements created by mankind serves well when considering material things; however, there are functions that involve only fundamental energy acting within singularities where time is not bound to mankind's measuring system that considers the consequence of traversing distance. The fundamental energy that powers all things is supplied through a singularity where interactions are instantaneous, where time is not involved. But upon being emitted (radiated) from a singularity, distance is involved. Fundamental energy emission (expansion, radiation) from its singularity travels at the speed of light (186,000 mps), a constant. So time becomes a factor in the rate of speed of fundamental energy travel.

Subconscious reasoning that takes place without conscious intervention (awareness) like vision, sleep, and sleep-associated functions are in the speed-of-light category, but conscious reasoning is delayed by the reasoning process, which provides an explanation for why time associated with some functions seems to vary as the degree of awareness (conscious involvement) varies. While learning how to control conscious involvement in normally subconscious matters may prove beneficial, the point being made is to recognize that a singularity of fundamental presence energy functions within each person. Reasoning leads to the conclusion that it is fundamental presence energy, not electric energy as now thought that is involved in brain and nerve activity. In other words, our beings are powered by the same source that powers creation. Bear in mind that electric energy is, in fact, fundamental energy insulated and isolated into an electric system of conductors. However, we are not connected to an electric power source through wires.

The point being made is that there is a singularity at the center, the core, of each and every assembly of matter, and it is through that singularity center that power to sustain the functioning of fundamental particles of which all bodies are made is supplied. The sustaining of fundamental particles must be recognized independently of the processes involved in providing the power for the assemblies of fundamental particles to do work. In the case of living things, it is through the behavior of fundamental particles that fundamental energy is extracted from nutrients so the assembly can do work. So it is with all assemblies regardless of their configuration or function, and so it is that life is the traversing of distance.

Added Comments

Our body is an assembly of fundamental particles that are powered by presence energy as are any other assembly of matter. Living things, plant and animal, are categorized as living because we have organs that provide added capability, such as growth, reproduction, mobility, and intelligence, but we are still powered by presence energy delivered through the fundamental particles of which it is made. In other words, it is presence energy that gives life.

The concept that a divine presence is in and sustains living things is not entirely new, but identifying the mechanics by which that happens is new. Presence energy powers the behavior of all things, including inorganic atoms and molecules, and it is the medium involved in vision.

BLOG NINE

Petroleum from Carbon Dioxide

The abiogenic petroleum creation process is driven by the force of gravity acting upon the proton-generated forces that bind electrons, protons, and neutrons to form atoms, molecules, and thereby all matter and upon the resistance of physical matter to compaction. Knowledge of the fundamentals of creation physics and interaction of gravity and the proton-generated binding forces is prerequisite to understanding the abiogenic petroleum creation process, which requires recognition that current understanding of fundamental physics principles is definitively impaired.

The Gravity-Producing Process

Begin by accepting that everything physical in the universe has been produced through the assembly of precisely uniform particles of divine-in-nature (meaning no discernible physical presence) material and that, in their fundamental state, possess only one behavioral characteristic—namely each particle has one unit of affinity (attraction power) for the other. As a consequence, the entire quantity of fundamental particles exists at a point, and while that assembly of particles presents no physical presence, it constitutes a singularity point that possesses an affinity or attraction potential equal to the number of fundamental particles.

Now accept that when those fundamental particles are energized, as with a charge comparable to the charge known as background radiation, each particle then carries a like charge. These charges create a repelling force between particles whereupon the particles are dispersed ubiquitously with space, distance, time, and gravity created as a consequence. Recognize that since the number of particles of that assembly remains unchanged, the affinity attraction potential of its singularity point, its center of gravity, likewise remains unchanged. Then recognize that the interaction of the attracting and repelling forces acting between individual particles results in configuration of the expanded assembly in the form of a sphere with the singularity, the center of gravity, at the center point. Also recognize that with the attraction potential of the singularity now radiated to fill the volume of its spherical form, the density and, therefore, the intensity of the radiated potential is distributed inversely proportional to the square of the distance as evidenced by the formula for determining the volume of a sphere; therefore, the intensity (strength) and the power to produce force is reduced in proportion to the square of the radial distance from the center of gravity.

While this is, in essence, the gravity or gravitational attraction-producing process, it must be recognized that the difference of potential between the singularity and each individual fundamental particle establishes an infinitely elastic pulling or attracting force on each individual particle independent of any other interparticle binding or resistive forces. In other words, the force of gravity produced by a singularity can cause the fracturing of inter-atom and inter-molecule *proton-generated* bonds depending upon the strength of opposing forces. Also, while gravity may cause assemblies of particles to compact and produce pressure upon underlying matter, the principal force is attraction of fundamental particles by the singularity point of the involved assembly.

The Proton-Generated Electron-Attraction Process

In essence, protons acquire a singularity of electron-charge-attracting energy at the time of their creation through transformation from a neutron. It is that proton-generated affinity attraction for an electron charge that produces the binding force involved in making atoms, molecules, and thereby all material things. The relationship between protons and electron charge is 1,839 to 1, the power of which is reduced with distance over which it is acting; therefore, the binding force produced is relatively easily fractured as occurs when electrons are caused to be moved from one atom or molecule to another, which is fundamental in electrical and chemical actions. It is the energy released when atom-to-electron bonds are broken that produces the effect called heat, and because that released energy is attracted by the proton-generated attraction singularities of atoms, electron-attracting strength is committed to bonding the heat-producing energy so it is not available to attract atoms. In other words, the acquisition of heat weakens the binding power of atoms and molecules since only residual strength not committed to binding heat energy is available to form atom-to-electron bonds. Furthermore, the atom-to-electron binding strength of the various atoms also varies depending upon their physical alignment. It is the fragility of the proton bonds that is of interest in discussing the abiogenic petroleum-creation process.

The Dalton and Avogadro Gas Laws

Dalton's law says that the total pressure of a mixture of gases (as the atmosphere for example) is the sum of the pressures produces by the individual gases.

Avogadro's law says that at the designated standard pressure and temperature, 22.4 liters of any gas contains 6.02×10^{23} molecules of that gas, and the weight of that quantity of molecules (designated as the gram-molecular weight) is in proportion to the atomic weight of the molecules under consideration. In other words, an oxygen molecule with an atomic weight of thirty-two weighs thirty-two grams. Unfortunately, Avogadro had concluded that protons and neutrons were the elementary particle of mass, which is now shown to be in error.

A priori then says Dalton's law is derived by a behavior that reacts to the gram-molecular weight of molecules and thereby reflects the composition of the molecules that, in turn, reflects the magnitude of the proton attraction and gravitational attraction charges. For example, the O2 molecule with sixteen protons and sixteen neutrons has an electron attraction power of sixteen units (one per proton) plus a gravitational attraction of thirty-two units.

Recognize that the strength of the proton-attraction charge radiated by molecules is element-specific and results in the establishment of attraction between molecule singularity point and electrons, including those of other molecules. In each case, the magnitude of the force produced is inversely proportional to the square of the distance over which it is acting. Also recognize that the gravitational attraction charge acting out of the singularity point of each molecule establishes an attraction force upon the fundamental particles of other molecules in the same manner. Then recognize that when dealing with matter in a vapor form, the combination of repelling and attracting forces bind the like element-specific molecules into an isolated lattice or colloidal-like structured assembly as evidenced by the familiar assembly of H2O molecules that are called clouds.

Finally, recognize that the gravitational attraction of the earth acting upon each fundamental particle in atoms and molecules produces the effect called weight with the pressure or force applied a function of the density of the concerned assembly and the distance from the earth's singularity point.

Partners of the Process

Water (H_2O) and carbon dioxide (CO_2) molecules are residue in the combustion of carbon-based petroleum products wherein oxygen is a catalyst for their behavior. Hydrogen (H) and carbon (C) are the principle constituents of petroleum that exists essentially as CH_4 molecules. The point being made is that the carbon and hydrogen atoms are sufficiently resistant to the effects of heat and pressure to survive combustion. There is a reason.

Hydrogen. The nucleus of a hydrogen atom is a proton, which is a fundamental component of all atoms and consequently all matter. Protons are the generator of the proton charge that produces the electron-charge-attraction force that binds the formation of atoms and molecules. It is a charge that is fundamental to all electromagnetic and chemical processes, including heat and light. While it is the generator of the charge that produces heat and it is the charge that attracts the energy residue of broken bonds that produces the effect called heat, the integrity of the proton is not affected by heat. Therefore, protons are indestructible as long as the gravitational relationship with the singularity of the universe is maintained. In other words, there will be protons even if all other matter has disintegrated.

Carbon. The physical arrangement of carbon atoms results in the development of very powerful bonding of its six neutrons, six protons, and six electron charges, and that gives carbon atoms a unique ability to resist the effects of heat. Carbon atoms are so compact that there is virtually no distance between the atom nucleus and its electron charges so that nearly all its electron-attracting power is committed to binding its own electrons with very little strength lost because of the inverse square rule. Therefore, very little excess electron-attracting power remains with which to attract the residue energy that produces the effect called heat. That makes carbon atoms second only to protons in their ability to resist the effect of heat.

Oxygen. Oxygen atoms are more loosely assembled so that the greater space (distance) between their nucleus and their eight electron charges, which means a significant portion of the electron-attracting power of their singularity remains with which to bind heat-producing energy and the electron charges of other atoms. It is that behavioral characteristic that causes oxygen atoms to be very chemically active so that oxygen atoms exist normally only in combination with other atoms. The oxygen that is a component of the earth's atmosphere exists as a two-atom molecule (O2) where the electron-attraction energy of the individual atoms is committed to binding each other and as an O2 assembly is considerably less chemically active. However, when subjected to sufficient heat as occurs in the process of combustion, the O2 bonds are fractured, freeing the very chemically active fundamental oxygen atoms.

Water. The fundamental water molecule is an assembly produced when the electron-attracting power of an oxygen atom binds the electrons of two hydrogen atoms. The hydrogen atom is the smallest; therefore, its power to produce binding (whether the electron attraction or the gravitational attraction types) is not significantly diminished as a result of the inverse square of the distance rule. Bear in mind that bonds established by gravitational attraction are infinitely elastic and cannot be broken, plus the power of gravitational attraction is not affected by heat. Then bear in mind that at the distances involved, gravitational attraction is a significant factor in the binding of H2O molecules.

In the process of forming the fundamental H2O molecule, the oxygen atom attracts the electron charge of the two hydrogen atoms, thereby stretching the distance and thereby reducing the amount of energy applied in forming their bond with the electron charges of the oxygen atom, which leaves the hydrogen atoms with a residual electron-attraction power or positive charge. At the same time, in attracting the two hydrogen electrons, the oxygen atom develops a deficit or negative electron-attraction charge. Therefore, the assembly of H2O molecules produces a lattice or colloidal-like structure and unique behavioral characteristics.

While gravitational attraction between molecules plays a significant role in the formation of H2O molecular assemblies, gravitational attraction of the earth (gravity) plays a most significant role in their behavior. Bear in mind that the earth's singularity point possesses one unit of affinity attraction for each of the individual fundamental particles of which the earth's assembly is made but that the strength of affinity force applied to a fundamental particle is inversely proportional to the distance from the earth's singularity. Progressive compression is a consequence of matter being pulled by gravitational attraction toward a singularity point, and increased density is a consequence of compression. At the same time, the strength of the gravitational attraction being applied increases as the distance from the singularity decreases. Therefore, the earth's gravitational attraction has a most significant impact upon the behavior of H2O molecules.

Because H2O molecules in the liquid state are essentially incompressible, the density of water is essentially constant even to the depths of the ocean, which is not true for material in a gaseous state, such as CO2. It is important to recognize that atmospheric and ocean-depth-pressure measurements are, in fact, the measurement of the force of gravity being applied to a representative sample of the medium (usually measured in square inches or square millimeters) at a specific distance from the earth's singularity point. Therefore, the increase in pressure with increased depth in the ocean is not a function of material density but of the force of gravity. When the density of an object is greater than water, the object will be drawn to the ocean floor while objects with density less than water will rise to the surface. That contrast in behavior is important when considering the behavior of other objects immersed in water, such as carbon dioxide.

Carbon dioxide. The carbon dioxide molecule (CO2) is an assembly produced when two of the chemically active oxygen atoms bond with electron charges of a carbon atom under elevated temperature conditions, such as exists as a consequence of combustion. By exerting opposing forces, the carbon atom electrons being attracted by the two oxygen atoms are pulled to increase their distance from the carbon atom singularity, reducing the power exerted by the carbon atom singularity as a consequence of the inverse square rule. When that happens, less of the carbon atom's electron-attraction power is consumed, leaving a residue that is now available to attract residual heat, and the normally chemically inactive carbon atoms become aggressively active. In other words, as a consequence of the oxygen atom's behavior, the carbon dioxide molecule is converted into a powerful attractor of heat that saps the binding power of the CO2 molecule, allowing expansion and reduction in density.

It is that change in the behavior of the carbon atom that explains the behavior of CO2 molecules. It explains why CO2 is an aggressive absorber of the energy that produces the effect called heat and why the CO2 molecule density is so affected by temperature. Furthermore, the sudden change in the behavior of the carbon atom explains why CO2 changes directly from a solid (frozen) to a vapor state without existing as a liquid, and since CO2 exists in a vapor state at the temperature and pressure at the earth's surface, it explains why the density of CO2 is also pressure sensitive. Finally, it should be recognized that CO2 is an absorber, not a generator of heat, and therefore a vital link in the earth's temperature-regulating process that has apparently functioned reasonably well for the life of the earth.

The Petroleum-Producing Process

Begin by recognizing that the density of CO_2 molecules and thereby gravity's effect upon those molecules is a function of the prevailing mean temperature and that residue energy that produces the effect attributed to heat is attracted by the singularity that impresses the strongest electron-heat attraction. In other words, when CO_2 gives up heat, then the molecules increase in density, and the force of gravity has greater effect, which results in increased pressure that further increases the CO_2 molecule density. That compounding effect is most dramatically revealed when CO_2 molecules are drawn into the ocean by gravity, where the density of the water is not appreciably increased by the force of gravity. As a consequence, the CO_2 molecules remain in the gaseous state as they are drawn to the depths of the ocean.

Now recognize that the effect of gravity on CO_2 is not limited to waters of the ocean alone nor does it cease at the ocean floor. To the extent, the pulling action is not blocked by friction and the resistance of more physically endowed matter; denser material is more forcibly pulled by the earth's center of gravity even to the extent that intermolecular bonds may be fractured in the process, thereby releasing the energy employed in establishing and maintaining those bonds. The bonds fragmented by the force of gravity result in the release of residual energy that becomes involved in producing the effect called heat.

Further recognize that the power of gravity increases as the distance from the earth's singularity decreases so that a point is eventually reached where the mean temperature destroys all atoms and molecule assemblies except the most resistant carbon and hydrogen (protons) atoms. In the course of that process, the bonds of CO_2 molecules are fractured, the carbon atoms become essentially indestructible, and the oxygen atoms expand to do what oxygen atoms do. In a like manner, the bonds of H_2O molecules are fractured, the hydrogen atoms without an electron are indestructible, and the oxygen atoms expand and do what oxygen atoms do. That leaves carbon atoms and the nucleus of hydrogen atoms (protons) alone, exposed to very high-temperature environment. As the temperature of the carbon atom increases by attracting the residue electron-binding energy, the electron—or heat-attracting power of the carbon singularity increases to the point where the hydrogen atoms are capable of establishing bonds with the four electrons in the outer orbit of the carbon atom, resulting in the formation of a molecule consisting of one carbon and four hydrogen atoms. This molecule is called methane, the fundamental molecule in the formation of hydrocarbon compounds. At formation, methane molecules are low-density and begin their journey back up through the earth's crust where their configuration is altered through interaction with the various materials encountered.

Lessons to Be Learned

1. The formation of petroleum is a continuing, naturally occurring process.
2. Heat causes global warming, not CO_2, and heat is an effect of the energy residue from broken atom-to-electron-charge bonds.
3. When icebergs and the polar caps melt, they take up heat from their surrounding environment, causing an increase in CO_2 molecule density and their removal from the atmosphere.
4. It is the near-perfect vacuum of the space surrounding the earth and the earth's magnetic force field that blocks the escape of heat from the earth, not CO_2 and other greenhouse gases.
5. Logic derived a priori demands acknowledging the existence of a source for the energy that sustains the viability of the universe.
6. Creation physics is not complicated.

Blog Ten

Vision Lensing

I sit here in my laboratory, studying the behavior of physics associated with vision, at this time the physics associated with observed phenomena related to perceptions that the moon and sun appear larger at dawn than at noon. There is currently no plausible explanation for the perceived difference in size of the moon and sun.

Moon on the Horizon, courtesy NASA

A theory is forming that these phenomena constitute vision lensing due to gravity just as Einstein theorized regarding light lensing, except in this instance it is the gravity of the earth that produces the lensing effect. Vision lensing phenomenon is concluded to be caused by the effect described as

Lorentz force. Lorentz concluded that a force is applied to a charged energy particle traversing an electromagnetic force field.

The term *electromagnetic force field* is usually associated with magnetic force fields. Electromagnetic force is inherent in the lines of magnetic force produced by protons. A line of magnetic force is an assembly of both the attracting and the repelling component of presence energy, where it is the repelling component that is segregated to produce electricity. Therefore, it is the repelling charge (called an electron) of presence energy that produces electromagnetic force. It is the attraction component of presence energy that produces gravity force, and it is the attraction component of presence energy that is perceived by our eyes, and that is vision. Electromagnetic force and gravity force are both products of presence energy. The point being made is that a gravity force field will have a similar effect on traversing charged particles as is produced by electromagnetic force fields. Vision involves the transfer of energy charges from the object being viewed to the eyes. That established leads to recognition that vision traversing a gravitational force field will respond to the Lorentz force; it can be concluded that Lorentz force can bend vision and that telescoping (vision lensing) can be produced with an appropriate configuration.

It is important to realize that while vision and light transmission are separate phenomena, they both involve charged energy particles. It is also important to realize that the energy that powers gravity exists as a force field that surrounds the earth. The intensity, or strength, of the force field is progressively less with distance from the center of gravity, plus its force acts perpendicular to the periphery of the earth. Several factors contribute to the lensing effect. First, the line of vision to objects on the horizon traverses the force lines that project perpendicular to the periphery of the earth. Second, the line of vision traverses the denser (the stronger) part of the gravity force field. Third, the Lorentz force causes constriction of the traversing charge particles the same as happens with a conventional telescope. By constricting the energy particles coming from the viewed object, the field of view is reduced, and the object appears larger. That process is called lensing or, in this instance, vision lensing.

It is apparent that the phenomenon called light lensing posited by Einstein is not light lensing but vision lensing and is not in response to the force of gravity of the causing body but of the force field of the causing body. Recognize that the intensity and density of the earth's magnetic force field decreases with altitude (distance from the center of the earth) and that the line of sight to the horizon traverses the direction that force field is acting while the line of sight to an object that is higher in the sky parallels the force field direction. Lorentz law says that when charged particles traverse a force field, their trajectory is bent. In the cited cases, vision of the distant objects is magnified traversing the lines of force of the earth's magnetic force field but not when the view parallels the lines of force. Consideration of that evidence leads to the conclusion that phenomenon produces vision lensing as seen in the photos as well as the effect called gravitation light lensing postulated by Einstein.

Added Comments

The theory that space and time are warped and that light is bent by gravity is not supported by fundamental physics. The physics principles explained in chapter 6 show that most of the current theories and explanations concerning light and vision are defective. With recognition of the fundamental physics principles explained in this publication, the world of physics will be upset.

Analysis of this phenomenon by following where the facts lead leads to a new understanding with regard to vision. Previous analysis led to the conclusion that vision functioned because our eyes perceive presence energy. Vision does not rely upon the radiation of photons by the objects being viewed as explained by the current accepted theory, yet the theory explaining light and vision lensing does depend upon the transmission of energy particles from the objects being viewed to the eyes. That realization drives the conclusion that presence energy also exists as quantum particles similar to photons, which follows the process of creation.

As explained in chapter 3, fundamental particles are charged with an attraction component and a repelling component that exist as quantum particles. Protons transform 1,840 attraction component charges into a photon. Photon radiation is the instrument involved with light and heat transmission. Photons radiate at the speed of light. The presence energy attraction component particle is fundamental to all creation; it would have an absolute intensity 1,840th the intensity of a photon, and it needs a name. The name *graviton* seems to fit, so it will be applied.

Accepting gravitons as the carriers of the energy involved with vision as well as gravity requires recognition that the graviton radiation process differs from the photon radiation process and requires further consideration.

BLOG ELEVEN

Supernova, Neutron Stars, Black Holes

In chapter 3 of my book manuscript pending publication, an explanation is provided on the process involved in the transformation of a neutron into a proton. The process involves the self-constriction of fundamental particles of matter (neutrino or God particles) that possess both attraction and repulsion charges of energy (yin-yang by ancient Chinese culture) to the point that the attraction charge is pulled off the fundamental particles into the center of gravity of the neutron. The collection of attraction charges standing alone without a repelling charge forms a singularity, that is, exists at a point. The repelling charges stripped from the fundamental particles surround the singularity, and the opposing force causes the repelling charges to spiral around the singularity core in the manner of an electric motor. The singularity emits lines of magnetic force that establish a magnetic force field. Stars are formed in the same manner except that with protons as the building blocks, the lines of magnetic force are significantly more powerful than those produced by protons. The process is powered through the presence-energy charge that powers gravitational attraction that powers creation.

My blog "Fueling the Sun" explains the processes involved within a star. A star is normally the center of a system of planets or mass that produces an unimaginably strong center of gravity. Being principally made of protons, each star has a center of proton power in addition to and acting independent of its center of gravity. The center of proton power attracts the residue of illumination, which is the energy that produces heat. The magnetic force field confines heat energy taken up by the singularity, and the retention of photon residue energy weakens the intensity of the star's singularity. If and when the expansion force produced by the retained heat energy exceeds the confining power of the magnetic force field, the star explodes as a supernova.

The process of dying involves a progression of steps that takes time. For a time, the involved protons, unable to establish intermolecular bonding, would continue to emit lines of magnetic force. As a consequence, the level of magnetic force, electromagnetic radiation, and rate of spin would increase as observed with pulsars. That process would gradually diminish as the protons were rendered impotent neutrons that have no singularity center. Eventually the dying star would be relegated to an intensely self-compacted mass of dead fundamental particles with an intense center of gravity and unable to produce lines of magnetic force and thereby light and heat, a black hole.

BLOG TWELVE

Mother Nature

Inferences of the creator as a mother go back to the Greeks in the twelfth and thirteenth centuries BC and by native North and South Americans before the time of European conquerors. Since the term infers giving birth and nurturing, it was a very appropriate descriptive term. Use of the term *father of creation* came into use as part of the Hebrew beliefs at the time of Abraham when males dominated the world and leaders were called lords. The point being made is that regardless of whether designated as male or female, the designation acknowledges belief in a creator even though the creator would not have been a human or an animal. Perhaps Mary was identified as the mother of God in an effort to correct a misinterpretation.

The foregoing rationalization shows how inconsequentially the term used to recognize creation and debate over the matter detracts from the real point. What exists now is the product of an event that produced all that now exists, which has been nurtured and through progressive development beginning with a fundamental form of energy was made into what now exists. Man, endowed with sophisticated intelligence, is able to study and gain understanding of the processes involved with the creation of what exists.

The business of studying what has been created is called physics, and therein rests the dilemma. Other than to draw conclusions a priori, what preceded the initial event cannot be understood through study of subsequent processes and a priori conclusions, which are beliefs, and beliefs are formed by our mind, which leaves us to acknowledge creation because of what exists. Some may choose not to believe the creator is God but refuse to acknowledge creation or that the expenditure of energy and the application of intelligence was a requisite. It is not an option to believe or not believe.

Similarly, since creation entails the production of what exists from elemental components, debate over evolution is meaningless. Progression of development is undeniably inherent in creation that begins with elemental components. Many question or are unsure that intelligence was involved in creation. Even in a simple decision of whether energy should attract or repel, that energy must know how, and that requires the application of intelligence. Even if it is assumed that the ability to choose is inherent (built in), then it must be recognized that intelligence was required to produce that inherent capability.

The same rationality applies to the claim of natural selection. To choose requires the application of intelligence, which could not exist without provisions with which to reason intelligently. To say natural selection is to say Mother Nature, which is to say the thing responsible for creation, which means an intelligent Mother Nature. Therefore, why not agree that an intelligent Mother Nature is the creator? Those who so choose to add belief in deity can take it from there. End of feud.

BLOG THIRTEEN

The Power in Believing

Psychology is intertwined with and subconsciously plays a significant role in all aspects of an individual's reasoning and behavior from birth until death. There is evidence of this in all forms of animal and, to some degree, plant life. Beginning at conception, behavior is controlled through DNA and other embedded instincts. The faculties needed to guide conscious thought and behavior are not available until a brain develops following which time we begin to learn and react to what has been learned. Mothers, especially, or the people filling that role, are aware of and significantly contribute to the learning process of infants and children in the course of nurturing and caring for their children's needs. Teaching love and basic right and wrong are foremost, which are in turn influenced by the beliefs of the mothering parent.

Those are the formative years when basic behavior influencing psychological beliefs is established. Psychological beliefs act to influence behavior, including reasoning, primarily subconsciously, almost instinctively. But beliefs are subject to being changed or altered, and a most critical period for the altering of mother-instilled beliefs occurs during adolescence; this period coincides with the time children are turned over to an outside system for education. Therein rests a major hidden problem.

Historically man has sought to understand creation, and the beliefs relative to the understanding of that phenomena is shown to have a significant influence on behavior. Religious beliefs and practices and civil laws have been established as a means of influencing behavior vis-à-vis interpretation of creation phenomena. As detailed by this writing, mass confusion and a long-standing feud prevails across society as a consequence of misunderstanding and misinterpretation of the processes involved in creation, which has and continues to have a deleterious effect upon beliefs and thereby behavior.

It is in the course of adolescent weaning from being mothered to self-controlled freewill that maturing children are exposed to the contrary beliefs vis-à-vis the behavior influencing creation phenomena. Speaking from personal experience as well as observation over a considerable lifetime, the impact of those conflicting beliefs relative to creation upon those passing through that period in life is enormous. The harm, the damage, is not just that misunderstanding interferes with technological advancement; the most damage is the result of confused behavior, including the perceived need for mind-altering drugs and medication that is driven by conflicting misinformation regarding this important belief-forming phenomenon that confronts adolescents.

Both sides in the feud profess strongly their interpretation and understanding of creation phenomena is correct and important, albeit not for the same reasons. The concern of the religious community is for the impact on human behavior to which they believe committed to correct. The opposing community of physicists and academia make their stand relative the creation phenomena principally in response to an unproven and unprovable anticreation, anticreator belief that was adopted, perhaps unwittingly, as an absolute law in physics. As a consequence, the problem stemming from the confusion of beliefs that confronts adolescents grows worse as civilization advances.

As in most feuds, contributing defective beliefs exist on both sides; as in most feuds, realignment of beliefs, rather than making matters worse, awakens enlightenment for each other; as in most feuds, the cause is hidden by narrow-minded, self-centered, and self-fulfilling missions. With recognition and awareness, the problem would self-destruct, but that cannot be achieved without a platform and a bullhorn to reach those in the feuding communities willing to champion reconciliation. A dilemma. Will it garner divine intervention?

Blog Fourteen

Black Holes / Dark Matter Explained

It takes a proton-electron bond to produce light and heat. Light and heat are effects produced by the energy of proton-electron bonds released when proton-electron bonds are broken as happens, for example, when electricity flows through the filament of a lightbulb or as a consequence of combustion.

Proton-electron bonds are produced by protons. An assembly of elementary particles (subprotons) is incapable of producing proton-electron bonds so are incapable of producing light and heat. Yet an assembly of subproton elementary particles possesses interparticle affinity (gravitational attraction) that mark the existence of mass.

Therefore, an assembly of subproton elementary particles (and/or impotent protons) would exhibit gravitational attraction but would produce no light or energy presence glow, which is the definition of a black hole. It also explains dark matter that is unassembled elementary particles.

Blog Fifteen

Explaining Time Dilation

Providing an explanation for time dilation is like killing two birds with one stone. While providing the answer in plain language, the answer proves the fundamental position of creation physics that the energy that supplies power to even the fundamental particles of matter is in the form of minute packets or pixels (quanta) with absolute intensity (strength). The energy that powers creation is supplied through the center of gravity of fundamental particles (neutrinos) and, in turn, the center of gravity of each assembly of matter as, for example, the earth.

The intensity of the charge of energy packets increases as the number of assembled particles increases. The physical size of the energy packets decreases as the intensity increases. Those packets of energy (quanta) establish a standing charge that surrounds the assembly where the packets of energy diminish in intensity and increase in physical size with increase in distance from the center of gravity. Because the charge is inherent in the fundamental particles and therefore each assembly of fundamental particles, the charge exists as an inherent component of each assembly of particles; the charge acts instantaneously. However, the energy supplied to maintain the charge radiates from the center of gravity at the speed of light so that passage at a specific point produces sine wave-like fluctuations or wave patterns that reflects the intensity of the energy particles available at that point.

The energy pixels (quanta) kept charged by the almighty energy reservoir established by the creator's presence give the fundamental particles the power to attract and repel, which then in an assembly add in proportion to the number of fundamental particles in an assembly to create an energy center, a center of gravity, a singularity at the center of each assembly. The earth has a large number of fundamental particles; therefore, the energy particles radiated by the earth are strong, high-intensity, high-density, compact, and small in physical size. This is a continuing action with energy expended to attract and bind matter at a rate set by the intensity at the point of application.

Picture a cloud of minute spherical energy packets surrounding an assembly of matter (say the earth) with the physical size of the individual spheres increasing and intensity decreasing with distance from the center of the assembly (earth). Recognize that each independent assembly is accompanied by this phenomenon with the charge of each of the assembly's fundamental particles continuously sustained through their individual center point while the charge of a multiparticle assembly is maintained through the center point of the assembly. It is that charge that gives each particle and each assembly of particles the power to function, to behave.

In dealing with radiated energy, it is the intensity of the energy at the point of application that determines the force applied. With speed at which radiating energy acts constant, the intensity of the energy and, therefore, force available to do work is dependent upon the number of the charges and the intensity of those charges. The mechanisms used in measuring the passage of time react to the intensity and number of charge spheres available.

Mechanical clocks employ a spring-driven escapement mechanism to regulate the rate of time lapse where the strength of the spring is the controlling factor. The spring strength is set by the strength of the force of interparticle binding, which reacts to the energy intensity being supplied. At higher altitudes (distance from the earth's center of gravity) the energy supplied is lower in intensity; the spring is not as strong, so the escape mechanism reacts more slowly and the clock runs more slowly. Atomic clocks react to the sine wave (frequency) of the energy they are tuned to monitor. At higher altitudes, the energy spheres are larger; therefore, the sine wave is longer and the frequency lower (slower in terms of time), so time lapse is slower.

The same reasoning applies to time dilation when objects are traveling at high speed. However, not only is the distance from the center of gravity a factor, moving objects are traveling through the radiated energy field at high speed, therefore increasing the number to energy spheres to which they are exposed. Under those circumstances, both countering phenomena apply. High-speed travel increases the energy intensity because of exposure to an increased number of particles in a given period as well as an increase in sensed frequency (shorter wavelength), both of which are reflected as faster time lapses. (Examples of time dilation where light, radio, or radar transmissions are involved have not been addressed because those methods of detection are subject to misinterpretation because of time delays in attempting to establish simultaneity.)

There is nothing mysterious about time dilation when it is recognized that all things are powered by the energy supplied through the particles of matter elemental to creation. Indication is that the health of living things (metabolism) would be affected by the quality of the presence energy supply in some manner, but the impact may or may not follow the behavior of time-keeping devices. That determination is not within the scope of this consideration.

BLOG SIXTEEN

Nuclear Fusion Does Not Produce Energy

For forty-plus years and billions of dollars, the best minds in the world have been trying to extract energy by nuclear fusion reaction. The program and concept has proven a colossal failure and has proven that modern physics theories encompass equally colossal defects.

The idea that energy in abundance beyond imagination could be obtained through nuclear fusion developed after the neutron bomb proved successful. Neutron bombs employ an atomic fission bomb to compress hydrogen atoms (protons) so they will combine (fuse) to form helium atoms (a theory) and in the process release copious amounts of energy. Subsequent testing proved the hydrogen (neutron) bomb was considerably more powerful than an atomic (fission) bomb alone. Because the resulting reaction produced a more intense explosion (release of energy), it was concluded that the greater energy yield was the product of fusion. That conclusion (theory) was defective because it failed to recognize that the intense energy released by the hydrogen (neutron) bomb was not released by fusion of the hydrogen atoms but by the fission of the helium atoms produced by the fusion of hydrogen atoms.

The concentrated pressure created by the atomic explosion was sufficient to affect fusion of hydrogen atoms to make helium atoms. The force generated by the energy expended by the activating fission explosion went into creating the conditions needed so *the bonds* of the new helium atoms could be created. The prevailing heat and pressure caused failure (fission) of the newly created bonds, thereby releasing the energy that powered those bonds. Not only is the energy intensity greater in helium bonds, there are considerably more bonds in a helium atom so that when a helium atom "splits," fails, or fissions, the magnitude and intensity of the energy released is greater and more concentrated. In addition, the force produced by the activating explosion adds to the force produced by fission of the helium atoms, producing understandably a more powerful reaction.

There is a preliminary matter that begs clarification. There are two forms of energy involved in creation.

Type 1 is the *attraction* component of fundamental energy that provides the power to establish gravity or gravitational attraction. This energy powers the fundamental binding of all assemblies of matter and thereby establishes mass. The bonds formed by the attracting component of fundamental energy are not frangible. These bonds will stretch and diminish in strength as the distance from the source increases but never break. Their strength will fade to the point of being undetectable, but these bonds will not break nor fragment to release their powering energy. The energy that powers gravitational attraction cannot be diverted for any purpose other than its fundamental role that includes the assembly of neutrons and the powering of protons.

Type 2 are the lines of force emitted by protons that establish bonds between protons. The energy to form bonds is *supplied* through the fundamental particles and incorporated into the proton-created lines of force. In other words, the new bonds formed to fuse the hydrogen to make helium did not just happen. New bonds are formed with energy supplied through proton-assembled photons. It is that new added energy that is released when the lines of force binding the helium atoms are fragmented. The bonds formed with proton-generated lines of force are frangible and when fragmented, release photons that radiate from the point of release at the speed of light. Photons are particles of fundamental attracting energy that are assembled by protons and power illumination and heat.

In other words, the energy that powers heat and light and thereby combustion and explosion, has *one source*, the photons that are a component of the lines of force created by protons. The energy expended in producing an atomic or nuclear explosion, or any combustion explosion, is supplied to protons through the fundamental particles employed in making protons. The protons, in turn, assemble that energy into photons that are encapsulated in the lines of force produced by protons and are released when the lines of force are fragmented.

The point being made is that *fusion requires energy input*, and fusion does not involve the fragmenting of proton bonds, so it does not release energy. Fission involves the fragmenting of the lines of force that produce proton binding and so, when fragmented, allows release of the energy involved in binding.

Blog Seventeen

The Defect in Mass Defect

The term *mass defect* refers to an apparent difference in the mass of assembled atoms when compared to the mass of the individual constituent components used in making the atom. When mass defect is translated into equivalent energy, the mass defect appears to show a loss of energy as a consequence of assembly (fusion) wherein interacting lines of force produce the force that binds the components into an assembly. That observation led to the conclusion that fusion reactions release (generate) energy.

The defect in mass defect begins with the methods employed in establishing the atomic mass (weight) of elements as shown in the periodic table (developed by chemists) in conjunction with the practice of using nuclei as the fundamental particle of matter even though nuclei are not the fundamental particle of gravity and therefore mass. The current practice assumes that protons and neutrons are the fundamental particles and sets their mass as being one-twelfth that of the assembly designated as a carbon 12 atom. Since carbon 12 is an assembly, its mass includes the effects resulting from or involved in effecting its assembly, yet the mass assigned to protons and neutrons was not adjusted to compensate for any changes resulting from assembly before the mass of the constitute components was established. For example, the impact of the inverse squared rule on the carbon 12 measured mass. Consequently, the mass of the various elements shown on the periodic table are flawed, and that is the cause of mass defect.

A major factor not included is determining the mass of an assembly is the effect of heat. As a consequence of applying compaction force by which to induce fusion, heat is released from bonds that are broken in the process. Increased heat produces expansion, lower density, lower gravitational attraction force, and the conclusion of reduced mass.

As an aside, use of the fundamental particles of which protons, neutrons, and electrons are made as the fundamental particle in the periodic table would be less controversial especially since that is the fundamental particle of gravitational attraction. Basing mass on the number of fundamental particles rather than attempting to convert the number of particles to an energy equivalent would eliminate all the variations inherent in an assembly.

It is the flawed conclusion that the assembled mass is less than the mass calculated by adding the mass of the contributing components that leads to the flawed perception that mass defect is caused by the assembly of atoms and that fusion, which entails the formation of new higher-intensity bonds and translates into the release of energy as heat. That conclusion defies logic and gives indication that the issue is forced in the interest of preserving the conservation of energy law, the conspiracy.

In the interest of providing an explanation for the release of energy as heat in conjunction with reactions involving fusion, such as with nuclear bombs, tokamak reactors, and solar activity, consider that under the conditions prevailing, the newly created helium (in those cases) is reduced by fission wherein energy is released in the form of heat by the bonds fragmented. To add further clarification, recognize that the energy to form the new higher-intensity bonds involved in fusion is supplied through the fundamental particles of which protons (hydrogen nuclei) are made.

For those insisting on the need to show proof, recognize this fact. The principles outlined are proven every time nuclear fusion is demonstrated, whether in a bomb, in a tokamak, or the sun, which is more logical proof than now offered to support the theory of mass defect. If necessary, numbers could be developed to support the principles offered as they have been for mass defect.

Blog Eighteen

Higgs versus God Particles

The Higgs theory envisions an unexplained field of something that fills all space wherein Higgs boson particles are somehow created. What Higgs particles are or how many are made or alive is unspecified. The Higgs boson particles interact somehow with claimed elementary particles of matter (quarks and leptons like electrons and neutrinos) to give those particles mass without explaining what mass is or does. Somehow quarks that do not, or cannot, exist as stable stand-alone particles are assumed to be the elementary particles involved in the formation of matter. The Higgs theory depends upon gluons to somehow know how to bind the elementary quarks and leptons to make protons and neutrons. And you are expected to understand and believe that.

The fundamental physics theory envisions a universe filled with presence particles with neutrino characteristics that are charged and kept charged by a charge field produced by the existence of the presence particles, known to exist as cosmic background radiation. When charged, each presence particle (God particle) has the power to attract the attraction charge as well as to repel the repelling charge of other presence particles, the yin-yang forces of ancient Chinese culture. The attraction force is known as gravitational attraction, and the ability to attract is known as mass. With the power to attract and bond with other presence particles, they serve as the elementary particles of creation physics, in which capacity they are the carriers of the energy that powers creation as well as giving physical presence to matter. The fundamental physics theory leaves only one unexplained condition: What or from what is the presence that is fundamental to the physics of creation? Is there anything left unexplained?

If someone with the power would submit this argument for peer review by the Nobel Commission, maybe physicists could get back on track to understanding fundamental physics.

BLOG NINETEEN

Trinity Particles

Reasoned consideration for the physics indicated to be required of a particle of energy with which to build the universe where nothing previously existed gives rise to the conclusion that creation would start with a massive quantity of absolutely identical elementary particles consisting of three energy components of equal intensity each with a distinct behavior. The primary component would establish an attraction force (gravity) with the attraction charges of other like elementary particles. A second component would establish a repelling force with the repelling charge of other like elementary particles. A third component would establish an attraction (an affinity) force between the attraction and repelling energy charges.

The gravity force interacting with the repelling force gives assemblies of elementary particles a physical presence. The affinity force interacting with the attracting and repelling charges unites the trinity of forces into a composite fundamental energy particle. With one intensity unit of attracting and one intensity unit of repelling energy assembled by one intensity unit of affinity energy, the charges of each trinity particle are in self-contained equilibrium and are detectable only by the forces and charges they project. A trinity particle or assembly of trinity particles would have mass and project gravitational attraction. A trinity particle would project a repelling force to another trinity particle, which is discernable when assembled into a proton but is not discernable when assembled as a neutron. Depending upon the state of assembly, the affinity component charge may or may not be detectable but when isolated from the other components is discernable as an electrical charge. The described elementary energy particle, the trinity particle, establishes the basis for absolutes that have been recognized including: energy is supplied as packets of distinct intensity (quantum); the power of the force produced is infinite and can neither be created or destroyed but diminishes in power inversely proportional to the distance from the source squared; the intensity of the charge that powers the elementary trinity particles is constant; the trinity elementary particle is the carrier of the energy that powers the universe and all that is in the universe.

It is the *attraction component* of the trinity particle that powers the assembly of elementary particles to form neutrons and protons where the force of attraction translates into gravitational attraction giving mass to each elementary particle and all assemblies of elementary particles. It is the attraction component that is formed into photons by the behavior of protons that gives lines of magnetic force the power to constrict and bind other assemblies with protons. It is the attraction component and the affinity component involved in forming photons that conveys to photons the power to produce illumination, vision and heat.

It is the *repelling component* of the trinity particle that encapsulates the *attracting component* and opposes the force of the attraction component giving rise to the formation of matter with a physical presence. It is the repelling component that protons use to surround and encapsulate attraction charges (1,836 units) to make photons. It is the repelling component applied to lines of magnetic force that conveys the power to repel. It is the repelling component encapsulating lines of magnetic force that is fragmented releasing photons to radiate when bonds involved in the formation of atoms and molecules are broken. It is the individual repelling component charges stripped off lines of magnetic force in an electric generator or a chemical reaction that are known as electrons. It is the repelling component charges (electrons) flowing along a conductor that gives rise to the magnetic force field of electromagnets and electric motors. It is the repelling component flowing along a conductor that makes and breaks bonds with each atom as they pass, that releases photons to produce light and heat. It is the repelling component of inter atom bonds reacting to microwave radiation that rupture to release photons that results in the production of heat.

It is the *affinity component* of the trinity particle that causes the repelling component to bond to, to encapsulate, the attracting component that holds the trinity of energy charges to form elementary particles. *It is the affinity between the attracting and the repelling components of fundamental energy that is isolated on a system of insulated conductors that provides the power involved in electrical and electromagnetic energy.* It is the affinity component that is active in chemical reactions. It is the affinity component released when the bonds between protons, and other assemblies made of protons, are broken that supports vision and is seen as the glow that surrounds bond fragmenting reactions when the magnitude of released affinity energy is sufficient.

Recognition of the described trinity of components inherent in fundamental energy that powers elementary (trinity) particles opens the door to other unrecognized possibilities. Consider that the trinity of components are the same, albeit interacting differently (flavors). Consider that if the power of affinity between attraction and repelling component charges can drive electricity, technology could be developed to isolate, insulate and drive the attraction component charges, the charges that produce the effect called heat, in a similar manner. In that regard bear in mind that repelling charges expand and dissipate in intensity when isolated from attracting charges while attracting charges isolated from repelling charges contract into a singularity, a center of gravity.

It is postulated that *in the course of proton and neutron formation* the trinity nature of the fundamental energy maintained on elementary particles of which protons and neutrons are made produces assembly in an arrangement known as quarks. That being the case, assuming each quark is an assembly of 613 elementary particles, quarks would form bonds with 613 times the power of a single elementary particle. Therefore, the combination of forces involved in forming protons and neutrons would include the interaction of the three X613 quarks plus the interaction of the 1,839 elementary particles. This is offered as an explanation for what has been called the strong force that is released when the particle bonds of protons and neutrons are fragmented. It is also offered as an explanation of why quarks do not exist except within whole protons and neutrons. It is also postulated that the explanation offered regarding the behavior of elementary particles explains Dark matter and Black Holes.

EPILOGUE

I am not a physicist; I am a retired professional engineer. Other than in high school, I have no formal education in physics. I purchased a college-level physics textbook in about 1956. It is well-worn, with many highlights. I struggled to understand fundamental physics especially as it deals with electricity, heat, light, and gravity. I bought a college-level chemistry book in the hopes that would help. It just didn't make sense. My logical engineering mind-set could not accept the textbook and encyclopedia explanations of physics principles, and it carried over into my efforts to reconcile Christian religious beliefs.

Eventually a series of epiphany-class revelations opened my mind so that slowly, a problem began to show, and understanding developed. They transcended the abyss that separates the Judeo-Christian-Islamic religious beliefs and the beliefs of modern physics. I do not use the word *epiphany* lightly, but I cannot otherwise explain the origin of understanding on matters that were troubling me throughout life.

Realizing that the energy fundamental to creation is present in the flame of a candle is especially rewarding. That realization opened the door to understanding the processes involved in gravity, light, heat, electromagnetism, and the molecular binding forces. Since candles have been a part of religious traditions from the earliest days of organized worship, it gives rise to wonder about the origin of the practice. I hope you realized enlightenment from this book as I have come to realize by writing it.

Index